AF552905

Economic Zoology

NIPA® GENX ELECTRONIC RESOURCES & SOLUTIONS P. LTD.
New Delhi-110 034

About the Authors

Dr. V. B. Sakhare is Principal of Yogeshwari Mahavidyalaya, Ambajogai (Maharashtra). He has 26 years' experience as an outstanding teacher and researcher and has done pioneering work in the field of Reservoir Fisheries, Limnology and Fish Biology. Prof. Sakhare has successfully organized more than 7 conferences and workshops.

Prof. Sakhare is member of Board of Studies in Zoology of Dr. Babasaheb Ambedkar Marathwada University, Chhatrapati Sambhajinagar and Walchand College of Arts and Science, Solapur (Autonomous college). Prof. Sakhare has authored/edited more than 45 books. He is a recognized research guide of Dr. Babasaheb Ambedkar Marathwada University, and Solapur University, Solapur.

His exemplary qualifications and experience in the field of 'Inland Fisheries' has benefited to several students as well as fish farmers and produced 4 Doctoral candidates, as also led to publications to spread his knowledge across diaspora.

Dr. A.D. Chalak is working as Assistant Professor in Department of Zoology at Late Shankarrao Gutte Mahavidyalaya, Dharmapuri (Maharashtra). She obtained her Ph.D. from Dr. Babasaheb Ambedkar Marathwada University, Chhatrapati Sambhajinagar. She has more than seven years teaching and research experience. She is a recognized research guide of Dr. Babasaheb Ambedkar Marathwada University, Aurangabad. Dr. Chalak is also working as Associate Editor of an international journal 'Ecology and Fisheries' (ISSN 0974-6323). She has published more than 20 research papers and five books. She also organized 'Student Solar Ambassador Workshop' on 2nd October 2019.

Dr. B. Vasanthkumar is presently working as Professor and Head of the Department of Zoology at Sir M.V. Government Science Degree College, Bhadravathi (Karnataka). He is recipient of summer research fellowship of Indian Academy of Science for year 2014 and 2017-18.He has published more than 70 research papers in the national and international journals and more than 60 popular science articles in different aspects of ecology and environment. He has also published 12 text books for under graduate students of Karnataka University, Dharwad. He is co-author of reference books like 'Aquatic Ecosystem and its management' 'Applied Ecology' 'Advances in Aquatic Ecology (Volume 7&8)' and 'Emerging Trends in Fisheries and Aquaculture'. Dr.Vasanthkumar also guided four students for M.Phil.

Economic Zoology

V.B. Sakhare
Professor
Post Graduate Department of Zoology
Yogeshwari Mahavidyalaya,
Ambajogai, Maharashtra, India

A.D. Chalak
Assistant Professor
Department of Zoology
Late Shankarrao Gutte Mahavidyalaya
Dharmapuri, Maharashtra, India

B.Vasanthkumar
Professor and Head
Department of Zoology
Sir M.V. Government Science Degree College
Bhadravathi, Karnataka, India

NIPA® GENX ELECTRONIC RESOURCES & SOLUTIONS P. LTD.
New Delhi-110 034

NIPA® GENX ELECTRONIC
RESOURCES & SOLUTIONS P. LTD.

101,103, Vikas Surya Plaza, CU Block
L.S.C.Market, Pitam Pura, New Delhi-110 034
Ph : +91 11 27341616, 27341717, 27341718
E-mail: newindiapublishingagency@gmail.com
www: www.nipabooks.com

For customer assistance, please contact
Phone: + 91-11-27 34 17 17
Fax: + 91-11-27 34 16 16

Print ISBN: 978-93-58875-28-7

ebook ISBN: 978-93-58871-57-9

Composed and Designed by NIPA®.

Preface

'Economic Zoology' deals with the application of zoological knowledge for the benefit of mankind. It is a specialized branch of zoology which deals with the animal world that is associated with the economy, health and welfare of humans.

The present book 'Economic Zoology' contains more than 1400 multiple choice questions on various aspects of applied zoology like fish farming, prawn farming, Aquarium keeping, pearl culture Dairy farming, Lac culture, Vermitechnology, Apiculture, Sericulture, Poultry farming etc.

The present book been specially developed keeping in mind the requirements of students, exam-aspirants and other readers with academic as well as competitions' point of view. The book is especially useful for the aspirants of various competitive examinations where objective applied zoology forms an essential part of zoology forms as essential part of the examination. The main aim of the book is to present the specialized subject of applied zoology in an objective and reader -friendly manner to make the readers study and practice numerous questions on its various topics thoroughly. The book comprises a wide spectrum of question, covering all important topics which are frequently asked in various examinations.

The book will act as an efficient tool to test your knowledge of the subject and preparation of the exam. The present book is the outcome of sincere efforts made by authors since last 3 years to bring ample number of multiple-choice questions. It will definitely provide ready reference to the students, researchers and teachers and will cover state and central civil service examinations, JRF, SRF, NET, SET and other allied examinations as well.

The authors are grateful to friends, students and colleagues for their encouragement and cooperation in preparation of this book. We are also grateful to Managing Director of NIPA Genx Electronic Resources & Solutions Pvt Ltd, New Delhi for bringing up the book in an attractive way in short span of time.

We hope that this book shall prove useful to all concerned and we invite suggestions and comments for improvement, which will be taken care at the time of revision of the book.

Vishwas Sakhare
Ashwini Chalak
B. Vasanthkumar

Contents

1

Fish Farming

1. Indian Major carps belongs to which family?
   a) Cobitidae b) Cyprinidae
   c) Ariidae d) Clupeidae
2. Body of carps is covered by which scales?
   a) Placoid b) Ganoid
   c) Cycloid d) Ctenoid
3. Family cyprinidae includes......
   a) Catla catla b) Labeo rohita
   c) Cirrhina mrigala d) All of these
4. Selection criteria for cultivable fish includes ....
   a) It should have a fast growth rate
   b) It should be a hardy and resistant to disease
   c) Its flesh should be tasteful and with high nutritive value
   d) All of these
5. Why carps are suitable for culture?
   a) They can tolerate low oxygen content
   b) They have fast growth rate
   c) They feed on plankton,decaying weeds and debris
   d) All of these
6. Which is the fastest growing Indian major carp?
   a) Labeo rohita b) Cirrhina mrigala
   c) Catla catla d) Labeo gonius
7. .... is a prolific breeder and breeds easily in ponds all the year round.
   a) Catla catla b) Labei rohita
   c) Tilapia mossambica d) Both a & b

8. .... has upturned mouth.
   a) Catla catla
   b) Cirrhina mrigala
   c) Labeo bata
   d) Labeo fimbriatus
9. Which of the following is surface zooplankton feeder?
   a) Labeo rohita
   b) Cyprinus carpio
   c) Wallago attu
   d) Catla catla
10. ....attains maturity when two years old.
   a) Catla cattla
   b) Cyprinus carpio
   c) Oreochromis mossambicus
   d) All of these
11. ....feed mainly on organic matter, sand, mud and plankton.
   a) Labeo rohita
   b) Cirrhina mrigala
   c) Catla catla
   d) Labeo gonius
12. Which of the following fish is bottom feeder?
   a) Cirrhina mrigala
   b) Cyprinus carpio
   c) Both a & b
   d) Hypophthalmichthys molitrix
13. In which year silver carp was introduced in India?
   a) 1957
   b) 1959
   c) 1952
   d) 1969
14. Match the pairs:

a) Silver carp	i) Aristichthys molitrix
b) Grass carp	ii) Cyprinus carpio
c) Africian magur	iii) Pygocentrus natterei
d) Tilapia	iv) Oreochromis mossambicus
e) Common carp	v) Hypophthalmichthys molitrix
f) Bighead carp	vi) Clarias batrachus
g) Pirranha fish	vii) Ctenopharyngodon Idella

   a) a=v,b=vii,c=vi,d=iv,e=ii,f=i,g=iii
   b) a=vi,b=vii,c=v,d=iv,e=iii,f=i,g=ii
   c) a=v,b=vii,c=vi,d=iv,e=i,f=ii,g=iii
   d) a=v,b=vii,c=vi,d=iv,e=i,f=ii,g=iii

15. Grass carp is also known as ....
    a) Common carp
    b) Pirranha
    c) Whit Amur
    d) Koi
16. Which of the following is native to Africa and Middle East?
    a) Grass carp
    b) Tilapia
    c) Silver carp
    d) Common carp
17. ....belongs to the family cichlidae under order perciformes?
    a) Tilapia
    b) Cyprinus carpio
    c) Koi
    d) Big head carp
18. The species of the genera .... is mouth brooder?
    a) Sarotherodon
    b) Oreochromis
    c) Both a & b
    d) Cyprinus
19. Match the pairs:

a) Mozambique tilapia	i) Oreochromis niloticus
b) Blue tilapia	ii) Oreochromis hornorum
c) Nile tilapia	iii) Oreochromis Zilli
d) Zanzibar tilapia	iv) Oreochromis mossambicus
e) Red belly tilapia	v) Oreochromis aureus

    a) a=iv,b=ii,c=i,d=iii,e=v
    b) a=iv,b=v,c=ii,d=iii,e=i
    c) a-iv,b=v,c=iii,d=i,e=ii
    d) a=iv,b=v,c=i,d=ii,e=iii
20. Nile tilapia attains sexual maturity at an age of ....
    a) 10-12 months
    b) 5-6 months
    c) 3-4 months
    d) 9-10 months
21. In India, tilapia was introduced in year ....
    a) 1956
    b) 1978
    c) 1952
    d) 1953
22. In which year the Nile tilapia was introduced to India ?
    a) 1960
    b) 1955
    c) 1970
    d) 1980
23. Ctenopharyngodon Idella mainly feeds on ....
    a) Zooplankton
    b) Microplankton
    c) Aquatic vegetation
    d) Phytoplankton

24. In India, which fish is known as 'weed fish'?

a) Grass carp
b) Tilapia
c) Common carp
d) Bighead carp

25. Which of the following has three varieties?

a) Cyprinus carpio
b) Oreochromis mossambicus
c) Hypophthalmichthys molitrix
d) Aristichthys nobilis

26. In which year common carp was introduced in India?

a) 1936
b) 1939
c) 1949
d) 1952

27. In which year tench along with crusian carp were introduced in Nilgiri hills?

a) 1947
b) 1874
c) 1960
d) 1930

28. Which is the first exotic fish introduced in India?

a) Gourami
b) Tilapia
c) Tench
d) Crusian carp

29. .... is native to East Asia.

a) Bighead carp
b) Tilapia
c) Common carp
d) Both a & b

30. A major problem of pond cultured tilapia is ....

a) Excessive reproduction
b) Stunting of fish due to overcrowding
c) Both a & b
d) Requirement of more food and space

31. In which technique five or more different types of fish species are cultured together?

a) Composite fish farming
b) Polyculture
c) Both a & b
d) Integrated fish farming

32. .... is a technique for obtaining the maximum fish yield from a pond.

a) Composite fish culture
b) Monoculture
c) Monosex culture
d) All of these

33. The advantages of composite fish culture are ....
    a) Different fish species are cultured
    b) High amount of fish production can be obtained
    c) This method is economically feasible
    d) All of these

34. Which is the least managed form of fish farming in which little care is taken?
    a) Extensive fish farming system
    b) Intensive fish farming system
    c) Both a & b
    d) None of these

35. Which type of soil is suitable for fish farming?
    a) Loamy soil b) Clay loam bottom
    c) Dark coloured soil d) All of these

36. .... is unsuitable for fish pond.
    a) Rocky soil b) Sandy soil
    c) Porous soil d) All of these

37. Suitable soil pH for fish pond is ....
    a) 6.5 to 7.5 b) 9.5 to 11.5
    c) 4.2 to 5.2 d) 11.5 to 12.5

38. Soil with .... is ideal for fish production.
    a) Over 1% organic carbon
    b) Less than 0.5% organic carbon
    c) pH more than 10
    d) Sand and rocks

39. Soils with available nitrogen range .... are considered suitable for fish production.
    a) Below 30 mg N/100 g soil
    b) Between 20-35 mg N/100 g soil
    c) Above 50 mg N/100 g soil
    d) Between 5 -8 mg N/100 g soil.

40. Availability and release of soil phosphorous is mainly ....
    a) Oxygen dependent b) Hardness dependent
    c) pH dependent d) All of these
41. Water temperature favourable for healthy fish growth is ....
    a) 10-15 °C b) 35-40 °C
    c) 12-17°C d) 15-35°C
42. The water pH range of .... is most suitable for pond fish culture.
    a) 6 to 9 b) 4 to 6.5
    c) 9.5 to 11.5 d) 11.5 to 12.5
43. The main sources of oxygen in the pond water are ....
    a) Diffusion of atmospheric oxygen
    b) Photosynthetic production by phytoplankton
    c) Both a & b
    d) Zooplankton respiration
44. .... refers to the total concentration of bases in water expressed in mg/l of calcium carbonate.
    a) Hardness b) Alkalinity
    c) pH d) Nitrogen
45. .... more than 40 mg/l is considered to be more productive.
    a) Dissolved oxygen b) Total alkalinity
    c) Free carbon dioxide d) Both b & c
46. Natural waters accumulate carbon dioxide through ....
    a) Respiration of aquatic plants b) Respiration of aquatic plants
    c) Both a & b d) None of these
47. Seperating the male tilapia and raising them separately is known as.........
    a) Monosex culture b) Monoculture
    c) Polyculture d) Both a & b
48. Which sex is preferred In monosex culture of tilapia?
    a) Female b) Male
    c) Both d) Transgenic tilapia
49. Match the pairs:
    a) Floating weed i) Hydrilla
    b) Emergent weed ii) Anabaena

c) Submerged weed iii) Pistia

d) Marginal weed iv) Vallisneria

e) Planktonic algae v) Typha

a) a=iii,b=iv,c=i,d=ii,e=v

b) a=iii,b=iv,c=i,d=v,e=ii

c) a=iii,b=ii,c=i,d=v,e=iv

d) a=iii,b=v,c=i,d=ii,e=iv

50. Which of the following is an emergent weed?

a) Otellia

b) Azolla

c) Ceratophyllum

d) Utricularia

51. Azolla and Eichhornia are ....

a) Emergent weeds

b) Floating weeds

c) Submerged weeds

d) Marginal weeds

52. Spirogyra and Pithophora are the examples of ....

a) Zooplankton

b) Diatoms

c) Filamentous algae

d) Rotifera

53. Selection criteria for chemical used to control aquatic weeds is ....

a) It should be cheap

b) It should be easily available

c) It should be nontoxic

d) All of these

54. Which method of weed control is most convenient and least expensive?

a) Chemical method

b) Mechanical method

c) Biological method

d) Both a & b

55. Which fish is used for biological control of weeds?

a) Grass carp

b) Tilapia

c) Common carp

d) All of these

56. .... is reported to consume 40-70% of its own weight of grass per day?

a) Common carp

b) Grass carp

c) Pila

d) Gold fish

57. Catfish culture is more profitable because ....

a) They have a good market demand.

b) They are more nutritious and of high medicinal value

c) Catfish culture involves low risk

d) All of these

58. Which of the following is an air breathing predatory fish?
    a) Heteropneustes b) Anabas
    c) Clarias d) All of these
59. Example of predatory fish is ....
    a) Wallago attu b) Ompak pabda
    c) Mystus gulio d) All of these
60. Channa spp are commonly known as ....
    a) Snake head fishes b) Forage fishes
    c) Goat fishes d) Mahseer
61. .... are used as food of catfishes.
    a) Foreign fishes b) Forage fishes
    c) Carps d) Trouts
62. Example of forage fish is ....
    a) Puntius sophore b) Oxygaster bacaila
    c) Both a & b d) Wallago attu
63. Aquatic insects are harmful in nursery ponds because ....
    a) Insect larvae feeds on spawn and early fry stage
    b) Insects compete with fish for food
    c) Both a & b
    d) None of these
64. Example of Hemiptera (Bugs) is ....
    a) Anisops b) Ranatra
    c) Belostoma d) All of these
65. Example of coleoptera (Beetles) is ....
    a) Cybister b) Gyrinus
    c) Both a & b d) Anisops
66. Method used for control of aquatic insects is ....
    a) Use of soap oil emulsion b) Use of kerosene
    c) Use of turpentine oil d) All of these
67. Insects such as water scorpion and water stick insects belongs to which order?
    a) Hemiptera b) Odonata
    c) Coleoptera d) None of these

68. Match the pairs:

a)	Back swimmer	i)	Ranatra elongate
b)	Water scorpion	ii)	Dineuties indicus
c)	Water stick insect	iii)	Notoonecta glauca
d)	Giant water bug	iv)	Crixa sp
e)	Water boatman	v)	Nepa species
f)	Dragon flynymph	vi)	Anax sp.
g)	Whirling beetles	vii)	Lithocerus inducus

a) a=iii,b=v,c=i,d=vii,e=iv,f=vi,g=ii
b) a=iii,b=v,c=ii,d=i,e=iv,f=vi,g=v
c) a=iii,b=v,c=iv,d=i,e=ii,f=vi,g=v
d) a=iii,b=vi,c=v,d=i,e=ii,f=

69. Match the pairs:

a)	Family Notonectidae	i)	Nepa species
b)	Famil Nepidae	ii)	Crixa sp.
c)	Family Belostomidae	iii)	Notonecta glauca
d)	Family Coirxidae	iv)	Cybister limbatus
e)	Family Dytiscidae	v)	Dineuties indicus
f)	Family Gyrinidae	vi)	Lithocerus indicus
g)	Family Aeshnidae	vii)	Anax sp

a) a=iii,b=ii,c=vi,d=i,e=iv,f=v,g=vii
b) a=iii,b=i,c=vi,d=ii,e=iv,f=v,g=vii
c) a=iii,b=ii,c=vi,d=i,e=iv,f=v,g=vii
d) a=iii,b=i,c=vi,d=ii,e=iv,f=vii,g=v

70. Who is known as 'Father of Biofloc'?

a) Yoram Avnimelech b) Rosee Britz
c) Charles Nomiz d) Kritz Frincisco

71. In India which fish is suitable for biofloc?

a) Tilapia b) Anabus
c) Pangasius d) All of these

72. Who is glorified as 'Father of Indian Blue Revolution'?

a) Dr.V.G.Jhingran b) Prof.Hiralal Chaudhary
c) M.S.Swaminathan d) Dr.S.Ayyapan

73. Fishes have ....
    a) Paired and unpaired fins
    b) Lateral line system
    c) Internal ear
    d) All of these
74. Cage culture was first started in ....
    a) India
    b) Cambodia
    c) Thailand
    d) Germany
75. How many types of cages are used in fish culture?
    a) 3
    b) 2
    c) 4
    d) 5
76. Cages are made-up of ....
    a) Floats
    b) Frames
    c) Sinkers
    d) All of these
77. Advantages of fish culture in cages are ....
    a) Economics of water
    b) Reduction in fish handling
    c) Reduces pressure on land resources
    d) All of these
78. Limitations of fish culture in cages are ....
    a) Difficult to apply when water surface is very rough
    b) Need for maintenance of high dissolved oxygen level
    c) High quality balanced feed is desirable
    d) All of these
79. Advantages of pen culture are ....
    a) Intensive utilization of available space
    b) Safety from predators
    c) Ease of harvest
    d) All of these
80. Disadvnatges of pen culture are ....
    a) High demand for oxygen and water flow
    b) Dependence on artificial feed
    c) Rapid spread of disease
    d) All of these

81. .... is defined as raising of fish in running water.
   a) Pen culture b) Cage culture
   c) Raceway culture d) Integrated fish farming

82. .... is an aquaculture system that incorporates the treatment and reuse of water with less than 10% of total water volume replaced per day.
   a) Raceway culture
   b) Recirculatory aquaculture system
   c) Intensive fish farming
   d) Both a & b

83. .... consist of mechanical and biological filtration components, pumps and holding tanks.
   a) Recirculatory Aquaculture system
   b) Pen culture
   c) Cage culture
   d) Raceway culture

84. Advantage of recirculatory aquaculture system is ....
   a) Extended durability of tanks and equipment
   b) Reduced dependency on antibiotics and therapeutants
   c) Risk reduction due to climatic factors, disesase and parasite impacts
   d) All of these

85. Disadvantage of recirculatory aquaculture system is ....
   a) Constant uninterrupted power supply is required.
   b) Capital cost of starting a recirculatory system is high
   c) Judicial use of water and land areas
   d) Both a & b

86. Species suitable for recirculatory aquaculture system are ....
   a) Lates calcarifer b) Etroplus suratensis
   c) Oncorhynchus mykiss d) All of these

87. Scientific name of the rainbow trout is ....
   a) Oncorhynchus mykiss b) Salmo trutta fario
   c) Tor khudree d) Tor putitora

88. Trout farming is common in ....
    a) Jammu and Kashmir
    b) Himachal Pradesh
    c) Uttarakhand
    d) All of these
89. The rainbow trout is ....
    a) Herbivorous
    b) Omnivorous
    c) Carnivorous
    d) Planktivorous
90. Paddy varieties suitable for rice-fish integrated farming are ....
    a) Tulsi
    b) Panidhan
    c) Rajarajan
    d) All of these
91. .... are known as 'biological aerators' in fish farming.
    a) Pigs
    b) Ducks
    c) Poultry birds
    d) Both b & c
92. .... are constructed in a plain area by digging soil.
    a) Embankment pond
    b) Dug out pond
    c) Bundhs
    d) All of these
93. .... is constructed in hilly areas.
    a) Dug out pond
    b) Embankment pond
    c) Stocking pond
    d) Both a & b
94. Pond dyke should be ....
    a) Compact
    b) Soild
    c) Leak proof
    d) All of these
95. The objectives of pond drying are ....
    a) To kill undesirable species and predatory fishes from the pond
    b) To help for liming
    c) Both a & b
    d) To add new water
96. Dike reconstruction is essential to ....
    a) Prevent the pond from over flowing during rainy season
    b) Prevent the breaking of the dike
    c) Maintain a certain slope of the pond
    d) All of these

97. In fish farming, periodic fish sampling at regular interval is important with a view to ....

a) Checking the health condition of the fish

b) Monitoring the growth rate of fish

c) Estimating survival and mortality of fish in the pond

d) All of these

98. .... is a mating system where both sexes have multiple partners during the breeding system.

a) Polygamous
b) Promiscuous
c) Gonochoristic
d) Both a & b

99. .... is a mating system in which an individual of one sex has multiple partners during the breeding system but individuals of the opposite sex have only one partner.

a) Promiscuous
b) Polygamous
c) Trogamous
d) Both a & b

100. .... means the female lays unfertilized eggs which are externally fertilized.

a) Viviparity
b) Oviparity
c) Ovoviviparity
d) Homozity

101. In Indian major carps .... has rough dorsal surface.

a) Pectoral fin of male
b) Pectoral fin of female
c) Pelvic fin of male
d) Pelvic fin of female

102. Belly in mature female carp is ....

a) Soft
b) Swollen
c) Bulging
d) All of these

103. Which of the following fish has greater fecundity?

a) Catla catla
b) Cirrhina mrigala
c) Labeo rohita
d) Tilapia mossambica

104. Which of the following is an example of GnRH analog available in the market?

a) Ovaprim
b) Ovatide
c) Ovapel
d) All of these

105. Which synthetic compound is manufactured by Syndel Laboratories, Canada?

 a) Ovatide b) Ovaprim
 c) Ovapel d) Stritz

106. .... is a synthetic compound launched by Hemmopharma ,Mumbai.

 a) Ovatide b) Ovaprim
 c) Ovapel d) Both a & b

107. .... is developed in Hungary?

 a) Ovatide b) Ovaprim
 c) Ovastriz d) Ovapel

108. .... is a combination of mammalian GnRH analoge and water soluble dopamine.

 a) Ovapel b) Ovatide
 c) Ovaprim d) None of these

109. .... consists of salmon gonadotropin releasing hormone analoge and domperidone.

 a) Ovaprim b) Ovapel
 c) Both a & b d) None of these

110. Linpe method is used for ....

 a) Disease management b) Water quality management
 c) Induced breeding d) Sex reversal technique

111. Which of the following exhibit parental care?

 a) Tilapia b) Silver carp
 c) Common carp d) Big head carp

112. .... denotes the egg laying capacity of a fish.

 a) Maturity b) Fecundity
 c) Sexuality d) Both a & b

113. In carp female ....

 a) The pectoral fin smaller and weak
 b) The genital aperture and vent are swollen and protruding with pinkish margins.
 c) The abdomen is conspicuously bulging and is soft to touch
 d) All of these

114. The process of egg laying by females and milt by males is known as ....
   a) Spawning
   b) Maturity
   c) Fecundity
   d) Epiboly

115. Factors required for breeding of indigenous carps in bundhs are ....
   a) Fast water current
   b) High dissolved oxygen
   c) Heavy monsoons that flood the shallow spawning areas
   d) All of these

116. Some organisms may accidentally become planktons and they are called ....
   a) Holoplankton
   b) Tychoplankton
   c) Hypoplankton
   d) Mesoplankton

117. Organims which complete their life cycle in drifting are known as ....
   a) Holoplankton
   b) Ultraplankton
   c) Meroplankton
   d) Hypoplankton

118. Which of the following are crustacean zooplankton?
   a) Daphinia
   b) Moina
   c) Cyclops
   d) All of these

119. Example of protozoan zooplankton is ....
   a) Diffugia
   b) Arcella
   c) Vorticella
   d) All of these

120. .... do not breed in confined water.
   a) Labeo rohita
   b) Cirrhina mrigala
   c) Catla catla
   d) All of these

121. Name of the stage in which testes reduced in size and shows resorption.
   a) Mature stage
   b) Spent
   c) Maturing
   d) Spawning

122. Name of the stage in which ovaries flaccid and shrunken?
   a) Spent
   b) Mature
   c) Maturing
   d) Both a & b

123. Which of the following spawn once a year?

a) Tilapia b) Common carp

c) Catla d) Both b & c

124. Maturity age of common carp is ....

a) Two years b) One year

c) Three years d) Four years

125. In carps, the rate of development and duration of incubation depend mainly on ....

a) Water temperature b) pH of water

c) Salinity of water d) All of these

126. .... increases the incubation period.

a) Lower temperature b) Higher temperature

c) pH between 7.5 to 8.5 d) Both b & c

127. The incubation period for eggs of Indian major carps is ....

a) 5 to 8 hours b) 78-80 hours

c) 48-50 hours d) 14-18 hours

128. Stripping method is used for which fish?

a) Trout b) Tilapia

c) Catla d) Rohu

129. Which method of stripping is more convenient and effective?

a) Dry method b) Wet method

c) Both a & b d) None of these

130. Eggs of .... are non floating and non adhesive.

a) Catla catla b) Labeo rohita

c) Cyprinus carpio d) Both a & b

131. Eggs of common carp are ....

a) Adhesive b) Yellow to light brown

c) Reddish d) Both a & b

132. Egg diameter of common carp is ....

a) 5 to 6.5 mm b) 3.8 to 4.5 mm

c) 1.0 to 2.0 mm d) 2.5 to 3.5 mm

133. Egg diameter of mrigal is ....
   a) 4.0 to 6.5 mm b) 1.0 to 2.5 mm
   c) 7.5 to 8.0 mm d) 2.5 to 3.0 mm
134. 'LINPE' method was developed by ....
   a) Lin, Hao-Ren; Peter, R.E. (1988)
   b) Lamark, Russel; Peter, R.E.(1990)
   c) Lewis, Rac; Peter, A.K. (2000)
   d) Lewis and Poxell (2000)
135. The main advantage of hatching hapa is ....
   a) Less cost b) Its large size
   c) Its shape d) Its porous nature
136. The main advantages of vertical jar hatchery are ....
   a) Very low water requirement
   b) It can be operated in a compact area
   c) Water temperature is maintained
   d) All of these
137. Disadvantages of vertical jar hatchery are ....
   a) It is made of glass, it can be easily damaged
   b) It is difficult to shift to different places
   c) Temperature control system is not provided
   d) All of these
138. .... is based on continuous flow of water for carp breeding.
   a) Hatching hapa b) Nursery pond
   c) Chinese hatchery d) Both a & b
139. Advantages of CIFE-D-81 modern hatchery are ....
   a) Easy for transportation and packing
   b) Water temperature can be controlled
   c) Difficult to break
   d) All of these
140. Salient features of CIFE-eco hatchery are ....
   a) Low cost of establishment
   b) Maintenance free long life due to FRP material
   c) More than 90% hatching success and survival of spawn
   d) All of these

141. Condition favourable for carp induced breeding is ....
   a) Cool, rainy days
   b) Fresh rainwater
   c) Flowing water
   d) All of these
142. Suitable water temperature range for carp egg hatching is ....
   a) 26 to 30 °C
   b) 35 to 40 °C
   c) 10 -20 °C
   d) 5-20 °C
143. Suitable range of dissolved oxygen for carp egg hatching is ....
   a) 2.5 to 3 mg/liter
   b) 5- 8 mg/liter
   c) 3-4 mg /liter
   d) 1.8 to 2.5 mg/l
144. In common carp, the average fecundity per kg body weight is ....
   a) 1,60,000 to 2,16,000 eggs
   b) 80,000 to 1,00,000 eggs
   c) 55,000 to 85000 eggs
   d) 400,000 to 4,50,000 eggs
145. The success of fish breeding mainly depends on ....
   a) Health of fish
   b) Maturity of fish
   c) Weight of fish
   d) Both a & b
146. Fry stage of catla has ....
   a) Conspicuous head
   b) Snout flat and pointed
   c) Gill reddish and body colour light green
   d) All of these
147 .Fry stage of Rohu has ....
   a) Distinct opercular outline
   b) Distinct black chromatophores visible behind the eyes on the head region
   c) Gently fimbriated lip margins
   d) All of these
148. In fry of .... Few black chromatophores present in the caudal region.
   a) Cirrhina mrigala
   b) Catla catla
   c) Labeo rohita
   d) Ctenopharyngodon idella
149. Fry stage of Cyprinus carpio is characterized by...........
   a) Dark eyes
   b) Formation of rudiments of pectoral fin
   c) Both a & b
   d) None of these

150. Fry stage of Heteropneustes fossilis has ....
    a) Short dorsal fin
    b) Long anal fin
    c) Spineless dorsal fin but clearly not visible
    d) All of these
151. Clarias batrachus is commonly known as ....
    a) Desimagur b) Walking catfish
    c) Magur d) All of these
152. Fry stage of Clarias batrachus has ....
    a) Long dorsal fin b) Long anal fin
    c) 4 pairs of barbels d) All of these
153. .... are fishes which spawn freely in column of water and the eggs usually float.
    a) Psammophils b) Pelagophils
    c) Ostracophils d) Phytophils
154. The division of zygote into smaller and smaller cellular units is known as ....
    a) Morula b) Cleavage
    c) Fertilization d) Fate map
155. .... is a chart showing the fate of each cell of an embryo?
    a) Fate map b) Blastula
    c) Morula d) Emboly
156. .... shows the position of various presumptive organs in the early embryo itself.
    a) Blastula b) Morula
    c) Fate map d) Emboly
157. .... is the extension of the blastodisc over the surface of the yolk of gastrula.
    a) Epiboly b) Emboly
    c) Fate map d) Both a & b
158. .... is a type of morphogenetic movement which refers to the inword migration of cells in the embryo.
    a) Emboly b) Epiboly
    c) Morula d) Gastrula

159. Which type of eggs found in teleostean fishes?

a) Telolecithal eggs b) Mesolecithal eggs
c) Isolecithal eggs d) None of these

160. .... are passively floating and drifting eggs.

a) Semibuoyant eggs b) Demersal eggs
c) Planktonic eggs d) Both b & c

161. .... neither float not perfectly settle on the bottom.

a) Planktonic eggs b) Semibuoyant eggs
c) Demersal eggs d) Both a & b

162. .... sink and settle near the bottom.

a) Demersal eggs b) Planktonic eggs
c) Semibuoyant eggs d) Meroblastic eggs

163. .... are fishes which lay their eggs in aquatic vegetation.

a) Lithophils b) Pelagophils
c) Phytophils d) Ostracophils

164. .... deposit their eggs inside a bivalve mollusc.

a) Pelagophils b) Ostracophils
c) Lithophils d) Phytophils

165. Before transportation, the fry is subjected to ....

a) Conditioning b) Buffering
c) Thinning d) All of these

166. The larvae emerging from the fertilized eggs after hatching is called ....

a) Fry b) Fingerling
c) Hatchling d) Both a & b

167. Which of the following is characterized by the presence of yolk sac?

a) Hatchling b) Fingerling
c) Fry d) Seed

168. After absorption of the yolk sac, hatchling is known as ....

a) Spawn b) Fingerling
c) Fry d) Yearling

169. Length of fry is about ....
   a) 10-20 cm
   b) 1-2 cm
   c) 8-9 cm
   d) 12-15 cm
170. How many days are required for the spawn to grow up to fry stage?
   a) 1-2 days
   b) 10-20 days
   c) 7-10 days
   d) 20-25 days
171. Length of fish fingerling is about ....
   a) 10-15 cm
   b) 1-2 cm
   c) 5-8 cm
   d) Above 50 cm
172. How many days are required for the fry to grow up to fingerling size?
   a) 80-90 days
   b) 30-60 days
   c) 1-2 weeks
   d) 5 days
173. In fish farming, spawn is stocked at the rate of ....
   a) 6-10 million/ha
   b) 2-3 million/ha
   c) 1-2 million/ha
   d) 15-25 million/ha
174. For long distance transportation of brood fish which anaesthetic agents are used?
   a) MS222
   b) 2-phenoxy ethanol
   c) Both a & b
   d) Bleaching powder
175. Time of seed stocking should be ....
   a) Cool morning
   b) Evening hours
   c) Both a or b
   d) None of these
176. Fish seed should not be stocked during ....
   a) Rain
   b) Strong sunny days
   c) Cool morning
   d) Both a & b
177. The pond where spawn is reared upto fry stage is known as ....
   a) Spawning pond
   b) Nursery pond
   c) Rearing pond
   d) Stocking pond
178. Factors responsible for causes of mortality in nursery are...............
   a) Sudden fluctuations in physico-chemical conditions of water
   b) Presence of predatory and weed fishes in pond
   c) Presence of aquatic insects
   d) All of these

179.Which method is used for removal of predatory and weed fishes from ponds?

a) Repeated drag netting
b) Use of hook and line
c) By poisoning the water
d) All of these

180. Magur is ....

a) Herbivorous
b) Omnivorous
c) Carnivorous
d) Planktivorous

181. Maturity age of Clarias batrachus is ....

a) Second year
b) First year
c) Third year
d) Fourth year

182. Mature male magur is identified by ....

a) Posterior end of the body elongated,tapering and slender
b) Abdominal portion smooth not swollen or bulging in position
c) Genital papilla present and long,conical in shape or pointed
d) All of these

183. Mature female magur is identified by ....

a) Colour of vent is light pink in colour
b) Genital papilla short, oval or button in shape
c) Pressure on vent results in release of mature ova
d) All of these

184. .... is a circular net having the shape of a large umbrella.

a) Hela Jal
b) Cast net
c) Bhesal Jal
d) Gill net

185. Cast net is also known as ....

a) Ghagria Jal
b) Cover pot
c) Gill net
d) Drag net

186. .... are wall like nets with floats attached to the head rope and sinkers to the foot rope.

a) Cast net
b) Bag net
c) Gill net
d) Basket trap

187. Which of the following is a type of gill net?

a) Drift net
b) Set net
c) Both a & b
d) Basket net

188. Set nets are also known as ....
   a) Stationary nets
   b) Drift nets
   c) Floating nets
   d) Both b & c
189. An ideal aquafarm site should have ....
   a) Good water holding capacity
   b) Effective and efficient water exchange system
   c) Good accessibility and marketing facilities
   d) All of these
190. Improper aquafarm site leads to ....
   a) Higher cost of construction
   b) Low production
   c) Higher cost of culture operation
   d) All of these
191. Critical problems associated with an improper aquafarm site are ....
   a) High turbidity
   b) Improper water exchange
   c) Excessive seepage
   d) All of these
192. Advantages of ground water sources are ....
   a) More dependable and more uniform
   b) Free from wild fish and predatory insects
   c) Constant water temperature throughout the year
   d) All of these
193. Disadvantages of ground water sources are ....
   a) Less dissolved oxygen
   b) High concentrations of dissolved iron and other metals
   c) Contain toxic gases like hydrogen sulphide, methane and carbon dioxide
   d) All of these
194. Pre-stocking management includes ....
   a) Eradication of weeds
   b) Eradication of predatory fishes
   c) Eradication of aquatic insects
   d) All of these
195. Post stocking management includes ....
   a) Supplementary feeding
   b) Harvesting
   c) Manuring
   d) Both a & b

196. Which poison is used as piscicide?
   a) Croton tiglium
   b) Millatia piscidia
   c) Derris trifoliata
   d) All of these
197. Overdose of cowdung results in
   a) Gill rot disease
   b) Oxygen deficiency
   c) Fish growth
   d) Both a & b
198. The different inorganic fertilizers used in fish pond are ....
   a) Phosphate fertilizers
   b) Nitrogen fertilizers
   c) Potassium fertilizers
   d) All of these
199. The inorganic nitrogenous fertilizers are ....
   a) Sodium nitrate
   b) Ammonium sulphate
   c) Ammonium nitrate
   d) All of these
200. Example of organic manure is ....
   a) Liquid manure
   b) Farmyard manure
   c) Green manuring
   d) All of these
201. Example of aquaculture agriculture integration is
   a) Rice-fish integrated farming
   b) Fodder-cattle-fish integration
   c) Rice-Azolla-fish integration
   d) All of these
202. Example of aquaculture-livestock integration is ....
   a) Fish-duck integrated farming
   b) Fish-poultry integrated farming
   c) Mulberry-silkworm-fish integration
   d) Both a & b
203. .... feed upon single type of food.
   a) Stenophagic fishes
   b) Monophagic fishes
   c) Euryphagic fishes
   d) Both a & b
204. .... feed upon certain selected kinds of food.
   a) Monophagic fishes
   b) Stenophagic fishes
   c) Euryphagic fishes
   d) Carnivorous fishes

205. .... feed upon a variety of different types of food.

a) Euryphagic fishes
b) Monophagic fishes
c) Monthermal fishes
d) Eurythermal fishes

206. .... are removed completely from the pond prior to stocking of spawn?

a) Aquatic insects
b) Frogs
c) Water snakes
d) All of these

207. .... is genetically improved rohu with about 17 % higher growth efficiency developed through selective breeding?

a) Jayanti Rohu
b) Nile Rohu
c) Labeo calbasu
d) Both a & b

208. 'Jayanti Rohu' was developed by ....

a) ICAR- CIFE
b) ICAR-CIFA
c) ICAR-CIBA
d) ICAR-CMFRI

209. GOI permitted introduction of GIFT through ....

a) Rajiv Gandhi Centre for Aquaculture
b) Central Institute of Freshwater Aquaculture
c) National Fisheries Development Board
d) National Bureau of Fish Genetic Resources

210. .... was launched to bring about Blue Revolution through sustainable and responsible development of fisheries sector in India.

a) Pradhan Mantri Matsya Sampada Yojana
b) Pradhan Mantri Aquaexpo-2022
c) Matsya-2022
d) Matsya India-2022

211. The aims and objectives of the Pradhan Mantri Matsya Yojana are ....

a) Harnessing of fisheries potential in a sustainable, responsible, inclusive and equitable manner.
b) Enhancing of fish production and productivity through expansion, intensification, diversification and productive utilization of land and water
c) Doubling fishers and fish farmers incomes and generation of employment
d) All of these

212. 'CIFAX' is used for control of ....
   a) Whirling disease
   b) Epizootic Ulcerative Syndrime
   c) Epithelioma papillosum
   d) Both a & b
213. Which of the following is a protozoan parasite?
   a) Ceratomyxa shasta b) Branchiomyces songuinis
   c) Pseudomonas psychrophile d) Hemophilus piscicum
214. Flexibacter columnaris causes ....
   a) Whitish patches on head region b) Lesion on gill
   c) Both a & b d) Dropsy
215. Ulceratuve necrosis of skin and epithelial hyperplasia is caused by ....
   a) Pseudomonas psychrophila b) Tricrococcus spp.
   c) Branchiomyces sanguinis d) Ceratomyxa shasta
216. Dropsy disease is treated by ....
   a) Chloromycetin b) Methylene blue
   c) Common salt d) Picric acid
217. Tail rot or fin rot is a ....
   a) Viral infection b) Bacterial infection
   c) Protozoan infection d) Fungal infection
218. Best treatment for tail rot is ....
   a) 1% solution of silver nitrate b) Dilute solution of acriflavine
   c) Both a & b d) Common salt solution
219. Furunculosis is caused by ....
   a) Aeromonas salmonicida b) Pseudomonas punctata
   c) Costia necatrix d) None of these
220. Eye disease in fish is caused by ....
   a) Aeromonas salmonicida b) Aeromonas liquefaciens
   c) Pseudomonas punctata d) Both a & b
221. Eye disease in fish is treated by ....
   a) Chloromycetion b) Potassium permanganate
   c) Both a & b d) Oxalic acid

222. The causative agent of fish pox is ....
    a) Herpesvirus cyprini
    b) Aeromonas salmonicida
    c) Aeromonas liquefaciens
    d) Both b & c
223. Columnaris disease is known as ....
    a) Cotton wool disease
    b) Dropsy
    c) Furunculosis
    d) White spot disease
224. Gill rot disease is also known as ....
    a) Branchimycosis
    b) Dropsy
    c) Whirling disease
    d) Slimness of skin
225. Which of the following found only on fish gills?
    a) Gyrodactylus
    b) Dactylogyrus
    c) I.multifilis
    d) All of these
226. .... has a paired eyes at anterior end.
    a) Gyrodactylus
    b) Dactylogyrus
    c) Costia necatrix
    d) Both b & c
227. .... is a copepod crustacean, which resembles a worm due to the loss of appendages.
    a) Fish louse
    b) Anchor worm
    c) Argulus
    d) Fish leech
228. Lernaea is also known as ....
    a) Anchor worm
    b) Ergasilus
    c) Argulus
    d) Fish leech
229. Ulcer disease in fish is caused by ....
    a) Aeromonas hydrophila
    b) Hemophilius piscium
    c) Gyrodactylus
    d) Both a & b
230. Cotton wool disease is caused by ....
    a) Cytophaga columnaris
    b) Flexibacter columnaris
    c) Both a & b
    d) Ergasilus

Answer Keys

1	b)	Cyprinidae
2	c)	Cycloid
3	d)	All of these
4	d)	All of these

5	d)	All of these
6	c)	Catla catla
7	c)	Tilapia mossambica
8	a)	Catla catla
9	d)	Catla catla
10	a)	Catla cattla
11	b)	Cirrhina mrigala
12	c)	Both a & b
13	b)	1959
14	a)	a=v,b=vii,c=vi,d=iv,e=ii,f=i,g=iii
15	c)	Whit Amur
16	b)	Tilapia
17	a)	Tilapia
18	c)	Both a & b
19	d)	a=iv,b=v,c=i,d=ii,e=iii
20	b)	5-6 months
21	c)	1952
22	c)	1970
23	c)	Aquatic vegetation
24	b)	Tilapia
25	a)	Cyprinus carpio
26	b)	1939
27	b)	1874
28	a)	Gourami
29	a)	Bighead carp
30	c)	Both a & b
31	c)	Both a & b
32	a)	Composite fish culture
33	d)	All of these
34	a)	Extensive fish farming system
35	d)	All of these
36	d)	All of these
37	a)	6.5 to 7.5
38	a)	Over 1% organic carbon
39	c)	Above 50 mg N/100 g soil
40	c)	pH dependent

41	d)	15-35°C
42	a)	6 to 9
43	c)	Both a & b
44	b)	Alkalinity
45	b)	Total alkalinity
46	c)	Both a & b
47	a)	Monosex culture
48	b)	Male
49	b)	a=iii,b=iv,c=i,d=v,e=ii
50	a)	Otellia
51	b)	Floating weeds
52	c)	Filamentous algae
53	d)	All of these
54	c)	Biological method
55	d)	All of these
56	b)	Grass carp
57	d)	All of these
58	d)	All of these
59	d)	All of these
60	a)	Snake head fishes
61	b)	Forage fishes
62	c)	Both a & b
63	c)	Both a & b
64	d)	All of these
65	c)	Both a & b
66	d)	All of these
67	a)	Hemiptera
68	a)	a=iii,b=v,c=i,d=vii,e=iv,f=vi,g=ii
69	b)	a=iii,b=i,c=vi,d=ii,e=iv,f=v,g=vii
70	a)	Yoram Avnimelech
71	d)	All of these
72	b)	Prof.Hiralal Chaudhary
73	d)	All of these
74	b)	Cambodia
75	c)	4
76	d)	All of these

77	d)	All of these
78	d)	All of these
79	d)	All of these
80	d)	All of these
81	c)	Raceway culture
82	b)	Recirculatory aquaculture system
83	a)	Recirculatory Aquaculture system
84	d)	All of these
85	d)	Both a & b
86	d)	All of these
87	a)	Oncorhynchus mykiss
88	d)	All of these
89	c)	Carnivorous
90	d)	All of these
91	b)	Ducks
92	b)	Dug out pond
93	b)	Embankment pond
94	d)	All of these
95	c)	Both a & b
96	d)	All of these
97	d)	All of these
98	b)	Promiscuous
99	b)	Polygamous
100	b)	Oviparity
101	a)	Pectoral fin of male
102	d)	All of these
103	c)	Labeo rohita
104	d)	All of these
105	b)	Ovaprim
106	a)	Ovatide
107	d)	Ovapel
108	a)	Ovapel
109	a)	Ovaprim
110	c)	Induced breeding
111	a)	Tilapia
112	b)	Fecundity

113	d)	All of these
114	a)	Spawning
115	d)	All of these
116	b)	Tychoplankton
117	a)	Holoplankton
118	d)	All of these
119	d)	All of these
120	d)	All of these
121	b)	Spent
122	a)	Spent
123	c)	Catla
124	b)	One year
125	a)	Water temperature
126	a)	Lower temperature
127	d)	14-18 hours
128	a)	Trout
129	a)	Dry method
130	d)	Both a & b
131	d)	Both a & b
132	c)	1.0 to 2.0 mm
133	a)	4.0 to 6.5 mm
134	a)	Lin,Hao-Ren;Peter,R.E. (1988)
135	a)	Less cost
136	d)	All of these
137	d)	All of these
138	c)	Chinese hatchery
139	d)	All of these
140	d)	All of these
141	d)	All of these
142	a)	26 to 30 °C
143	b)	5- 8 mg/liter
144	a)	1,60,000 to 2,16,000 eggs
145	d)	Both a & b
146	d)	All of these
147	d)	All of these
148	a)	Cirrhina mrigala

149	c)	Both a & b
150	d)	All of these
151	d)	All of these
152	d)	All of these
153	b)	Pelagophils
154	b)	Cleavage
155	a)	Fate map
156	c)	Fate map
157	a)	Epiboly
158	a)	Emboly
159	a)	Telolecithal eggs
160	c)	Planktonic eggs
161	b)	Semibuoyant eggs
162	a)	Demersal eggs
163	c)	Phytophils
164	b)	Ostracophils
165	a)	Conditioning
166	c)	Hatchling
167	a)	Hatchling
168	a)	Spawn
169	b)	1-2 cm
170	c)	7-10 days
171	a)	10-15 cm
172	b)	30-60 days
173	a)	6-10 million/ha
174	c)	Both a & b
175	c)	Both a or b
176	d)	Both a & b
177	b)	Nursery pond
178	d)	All of these
179	d)	All of these
180	c)	Carnivorous
181	b)	First year
182	d)	All of these
183	d)	All of these
184	b)	Cast net

185	a)	Ghagria Jal
186	c)	Gill net
187	c)	Both a & b
188	a)	Stationary nets
189	d)	All of these
190	d)	All of these
191	d)	All of these
192	d)	All of these
193	d)	All of these
194	d)	All of these
195	d)	Both a & b
196	d)	All of these
197	d)	Both a & b
198	d)	All of these
199	d)	All of these
200	d)	All of these
201	d)	All of these
202	d)	Both a & b
203	b)	Monophagic fishes
204	b)	Stenophagic fishes
205	a)	Euryphagic fishes
206	d)	All of these
207	a)	Jayanti Rohu
208	b)	ICAR-CIFA
209	a)	Rajiv Gandhi Centre for Aquaculture
210	a)	Pradhan Mantri Matsya Sampada Yojana
211	d)	All of these
212	b)	Epizootic Ulcerative Syndrime
213	a)	Ceratomyxa shasta
214	c)	Both a & b
215	a)	Pseudomonas psychrophila
216	a)	Chloromycetin
217	b)	Bacterial infection
218	c)	Both a & b
219	a)	Aeromonas salmonicida
220	b)	Aeromonas liquefaciens

221	c)	Both a & b
222	a)	Herpesvirus cyprini
223	a)	Cotton wool disease
224	a)	Branchimycosis
225	b)	Dactylogyrus
226	b)	Dactylogyrus
227	b)	Anchor worm
228	a)	Anchor worm
229	d)	Both a & b
230	c)	Both a & b

2

Freshwater Prawn Farming

1. To which family freshwater prawn belong?
   a) Arthropoda
   b) Crustacea
   c) Palaemonidae
   d) Malacostraca
2. Scientifc name of freshwater prawn is ....
   a) Penaeus monodon
   b) Scylla serrata
   c) Macrobrachium rosenbergii
   d) Metapenaeus monoceros
3. Freshwater prawn is also known as ....
   a) Giant river prawn
   b) Scampi
   c) Banana prawn
   d) Both a & b
4. Identification features of freshwater prawn are ....
   a) Presence of grey and longitudinal streaks of light and dark colour
   b) The largest pereiopod of adult male is dark blue.
   c) The eggs of berried female are yellow in colour
   d) All of these
5. Freshwater prawn is known as shellfish because ....
   a) It has a hard exoskeleton
   b) It has scales
   c) It lives is shell
   d) Both a & b
6. Prawn is ....
   a) Herbivorous
   b) Omnivorous
   c) Carnivorous
   d) Both b & c
7. Cephalothorax in prawn is formed by the union of ....
   a) Head and thorax
   b) Abdomen and head
   c) Abdomen and eyes
   d) Somites and head
8. The crustacean biramous appendages have a basal part known as ....
   a) Protopodite
   b) Mesothorax
   c) Epipodite
   d) Chelate leg

9. The sex of prawn can be identified externally by observing which appendages?

   a) II chelate leg b) I maxillipedes

   c) I abdominal appendages d) II non-chelate legs

10. Male prawn can be differentiated from female by presence of ....

    a) Appendix masculine in 2nd pleopod

    b) Appendix masculine and appendix interna in 2nd pleopod

    c) Appendix interna in typical appendage

    d) Absence of appendix masculine in 2nd pleopod.

11. How many segments present in cephalothorax region?

    a) 11 b) 15

    c) 13 d) 21

12. Number of walking legs in prawn are ....

    a) 4 pairs b) 5 pairs

    c) 2 pairs d) 6 pairs

13. A statocyst lies inside ....

    a) The basal segment or precoxa of each antennule

    b) 4 th chelate leg

    c) Lateral line system

    d) Optic nerve

14. Function of statocyst in prawn is ....

    a) The Organ of orientation and equilibrium

    b) The organ of swimming

    c) The organ of hearing

    d) Sound producing organ

15. In prawn .... is located at the base of eye stalk, secrete many hormones.

    a) Pineal gland b) Sinus gland

    c) Hypophysis d) Sinusoid gland

16. Prawn is ....

    a) Dioecious b) Nocturnal

    c) Monoecious d) Both a & b

17. In freshwater prawn ....
    a) Male is larger than female
    b) Male possesses a narrow abdomen than female
    c) In male, second chelate legs are longer,stronger and more spiny than in female
    d) All of these
18. In prawn fertilization is ....
    a) External b) Internal
    c) Both a & b d) None of these
19. Blood of prawn is ....
    a) Colourless b) Red coloured
    c) Yellow coloured d) Green coloured
20. The respiratory pigment in prawn is ....
    a) Haemoglobin b) Haemocyanin
    c) Echinochrome d) Haemoerythrine
21. Blood-vascular system of prawn is ....
    a) Open type b) Lacunar type
    c) Both a & b d) Closed type
22. Blood-vascular system of prawn is characterized by ....
    a) Absence of capillaries b) Presence of lacunae
    c) Both a & b d) Absence of pericardium
23. Podobranch is also known as ....
    a) Foot gill b) Joint gill
    c) Side gill d) Lateral gill
24. is attached to the coxa of an appendage.
    a) Podobranch b) Arthrobranch
    c) Pleurobranch d) Both a & b
25. Arthrobranch is also known as ....
    a) Lateral gill b) Foot gill
    c) Joint gill d) Side gill
26. .... is attached to the arthrodial membrane joining a limb with the body.
    a) Arthrobranch b) Podibranch
    c) Pleurobranch d) Pseudobranch

27. Pleurobranch is also known as ....
    a) Foot gill b) Joint gill
    c) Side gill d) Pseudogill
28. Number and disposition of respiratory organs of each gill chamber in prawn is represented in the form of ....
    a) Branchial basket b) Gill formula
    c) Branchial formula d) Sclerite formula
29. Gills in prawn are ....
    a) Phyllobranch type b) Pseudobranch typr
    c) Holobranch type d) Both a & b
30. The excretory system of adult prawn consists of ....
    a) Green glands b) Lateral ducts
    c) Renal sac d) All of these
31. .... is known as brain of prawn.
    a) Circum-oesophageal commissures
    b) Ventral thoracic ganglionic mass
    c) Supra-oesophageal ganglia
    d) Ventral nerve cord
32. Eye of prawn is made of a large number of visual elements called ....
    a) Nephridia b) Ospradium
    c) Ommatidia d) Retinal stars
33. .... are the seed prawns.
    a) Larvae b) Post larvae
    c) Juveniles d) All of these
34. .... in ponds act as hide outs for small prawns.
    a) Palm trees b) Bamboo poles
    c) Tapioca d) All of these
35. In prawn farming feed is given at the rate of .... of body weight in 3 days.
    a) 1 to 2% b) 2-3%
    c) 6-7% d) 5-10%
36. The production of prawn farming is ....
    a) 20-25t/ha/6 months b) 2 t/ha/6 months
    c) < 1t/ha/6 moths d) 45-50t/ha/6 months

37. In polyculture freshwater prawn is cultured with ....
    a) Grass carp b) Silver carp
    c) Tilapia d) All of these
38. In .... the pond is stocked twice in a year and two crops are made.
    a) Batch culture b) Pen culture
    c) Cage culture d) Extensive culture
39. Prawn culture with paddy is known as ....
    a) Integrated prawn culture b) Intensive prawn culture
    c) Semi-intensive prawn culture d) Both a & b
40. Pokkali culture is practiced in ....
    a) Kerala b) Tamil Nadu
    c) Karnataka d) Haryana
41. Optimum water temperature for freshwater prawn farming is ....
    a) 18-32^{0}C b) 10 -14^{0}C
    c) 5 to 10 ^{0}C d) 35-42^{0}C
42. The attainment of first maturity in *M.roesenbergii* is at ....
    a) 175 mm b) 110 mm
    c) 90 mm d) 250 mm
43. During .... delicate post-larvae (15-20 mm) procured from hatcheries and raised to juveniles (2-5 gm) in small earthen ponds.
    a) Grow out phase b) Nursery phase
    c) Stocking phase d) None of these
44. In .... juvenile prawns (2-5 gms) are harvested from nursery ponds and stocked in larger grow out ponds @ 3-4/m2.
    a) Grow out phase b) Nursery phase
    c) Middle phase d) Both b & c
45. .... has developed a viable hatchery technology for the production of high quality seed and for table size prawn production.
    a) National Fisheries Development Board
    b) Central Inland Fisheries Research Institute
    c) Central Institute of Brackish water aquaculture
    d) Central Institute of Freshwater Aquaculture

46. Berried females in prawn are also known as ....
    a) Ovigerous b) PL
    c) Spent d) None of these
47. Which is the causative agent of white spot disease ?
    a) White Spot Syndrome Virus (WSSV)
    b) Costia necatrix
    c) Myxobolus cerebralis
    d) Both a & b
48. .... targets the cuticular epidermis, stomach, gills and hepatopancreas.
    a) Macrobrachium muscle virus (MMV)
    b) White spot syndrome baculovirus (WSBV)
    c) Costia necatrix
    d) Myxobolus cerebralis
49. White tail disease is caused by ....
    a) Macrobrachium rosenbergii nodavirus
    b) Aeromonas
    c) Costia necatrix
    d) All of these
50. The clinical signs of white tail disease are ....
    a) Lethargy
    b) Opaqueness of the abdominal muscle
    c) Whitish muscle appearance
    d) All of these

Answer keys

1	c)	Palaemonidae
2	c)	Macrobrachium rosenbergii
3	d)	Both a & b
4	d)	All of these
5	a)	It has a hard exoskeleton
6	b)	Omnivorous
7	a)	Head and thorax
8	a)	Protopodite
9	a)	II chelate leg
10	b)	Appendix masculine and appendix interna in 2nd pleopod
11	c)	13

12	b)	5 pairs
13	a)	The basal segment or precoxa of each antennule
14	a)	The Organ of orientation and equilibrium
15	b)	Sinus gland
16	d)	Both a & b
17	d)	All of these
18	a)	External
19	a)	Colourless
20	b)	Haemocyanin
21	c)	Both a & b
22	c)	Both a & b
23	a)	Foot gill
24	a)	Podobranch
25	c)	Joint gill
26	a)	Arthrobranch
27	c)	Side gill
28	c)	Branchial formula
29	a)	Phyllobranch type
30	d)	All of these
31	c)	Supra-oesophageal ganglia
32	c)	Ommatidia
33	d)	All of these
34	a)	Palm trees
35	d)	5-10%
36	b)	2 t/ha/6 months
37	d)	All of these
38	a)	Batch culture
39	a)	Integrated prawn culture
40	a)	Kerala
41	a)	18-32^{0}C
42	a)	175 mm
43	b)	Nursery phase
44	a)	Grow out phase
45	d)	Central Institute of Freshwater Aquaculture
46	a)	Ovigerous
47	a)	White Spot Syndrome Virus (WSSV)
48	b)	White spot syndrome baculovirus (WSBV)
49	a)	Macrobrachium rosenbergii nodavirus
50	d)	All of these

3

Aquarium Keeping

1. The word aquarium for the first time was used by ....
   a) Philip Henry Gosse
   b) Anne Thynne
   c) Philip Rottz
   d) Britz Carles
2. Who built the first stable marine aquarium in 1846?
   a) P.H.Gosse
   b) Philip Rottz
   c) Britz Carles
   d) Anne Thynne
3. In 1853,the first large public aquarium was opened in ....
   a) Roman empire
   b) London zoo
   c) Victoria era
   d) None of these
4. Taraporewala aquarium came into existence in ....
   a) Year 1951
   b) Year1945
   c) Year 1974
   d) Year 1981
5. Taraporewala aquarium is located at ....
   a) Bangalore
   b) Mumbai
   c) Hyderabad
   d) Barrackpore
6. Tar and silicone sealant allowed the first all-glass aquaria made by ....
   a) Britz Carles
   b) Anne Thynne
   c) Martin Horowitz
   d) S.Rotla
7. Indigenuos ornamental fishes used in aquarium is ....
   a) Puntius sp.
   b) Boti asp.
   c) Goldfish
   d) Both a & b
8. The Indian marine ornamental fish trade is benefited by ....
   a) Induced breeding
   b) Estuarine collection
   c) Wild collection
   d) Backwaters

9. The creatures and organisms live on the surface and inside of the ....

   a) Live rock b) Coral reefs

   c) Sponges d) Anemones

10. Advantage of live rock is ....

    a) It helps maintain stability in a saltwater aquarium

    b) It aids in nitrogen cycle

    c) Porous material inside the live rock helps rid of aquarium nitrates

    d) All of these

11. Which of the following is suitable for Coldwater aquarium?

    a) Goldfish b) Zebra fish

    c) Koi d) All of these

12. .... is the battery sealing compound.

    a) Silicone sealant b) Bitumen

    c) Zinc oxide d) Red lead

13. Disadvantage of bitumen is ....

    a) It is hard sealing compound

    b) It creates problems at the time of leak repair

    c) Readymade melted bitumen is not available

    d) All of these

14. Cementing substance suitable for aquarium tank construction is ....

    a) Bitumen b) Silicone sealant

    c) Linseed putty d) Zinc oxide

15. .... is used to apply the sealant to join the glass panes together.

    a) Squeezing gun b) Squeeze cocker

    c) Nozzle d) All of these

16. Linseed putty is a mixture of ....

    a) Zinc oxide b) Red lead

    c) Linseed oil d) All of these

17. .... contains many species of fishes.

    a) Community tank b) Species aquarium

    c) Biotope aquarium d) Public aquarium

18. Barbs belongs to which family?
    a) Cyprinodontidae b) Cyprinidae
    c) Bagaridae d) Siluridae
19. Which of the following is an example of Barb?
    a) Barbus tetrazona b) Puntius titteye
    c) Barbus conchonius d) All of these
20. Loaches are commonly known as ....
    a) Botia b) Barb
    c) Guppies d) Swordfish
21. Loaches are native of ....
    a) India b) Thailand
    c) Pakistan d) All of these
22. Spiny eels belongs to family ....
    a) Anguilldae b) Mastacembelidae
    c) Anabantidae d) Notopteridae
23. Badis belongs to which family?
    a) Nandidae b) Chromatidae
    c) Cyprinidae d) Siluridae
24. Badis also known as ....
    a) Dwarf Gourami b) Dwarf chameleon
    c) Climbing perch d) Police fish
25. To which family Puffer fish belong?
    a) Tetradontidae b) Anguillidae
    c) Cyprinodontidae d) Anabantidae
26. Channa sp are known as ....
    a) Barbs b) Singhi
    c) Snakeheads d) Eels
27. Family Chanidae is also known as ....
    a) Ambassidae b) Glossidae
    c) Ctenodi d) Percoidi
28. Glassfish belongs to which family?
    a) Siluridae b) Clupeidae
    c) Ambassidae d) Poecillidae

29. Example of glassfish is ....
    a) Chanda ranga b) Chanda nama
    c) Both a & b d) Etroplus suratensis
30. Scientific name of 'Rainbow fish' is ....
    a) Poecilia reticulata b) Etroplus chuna
    c) Ambassis nama d) Ambassis ranga
31. .... is also known as 'millions fish'.
    a) Angel fish b) Guppy
    c) Discus fish d) Jelly fish
32. Which of the following is a member of family Poecillidae?
    a) Anguilla b) Rasbora
    c) Anabus d) Guppy
33. Swordtail belongs to which family?
    a) Poecilidae b) Mastacembelidae
    c) Gobidae d) Notopteridae
34. .... is a member of cichlid family?
    a) Ophiochephalus punctatus b) Notopterus chitala
    c) Astronotus ocellatus d) Cyprinus carpio
35. Siamese fighting fish is a member of family ....
    a) Osphronemidae b) Notopteridae
    c) Channidae d) Mastacembelidae
36. In .... males are very aggressive towards other males of their own kind and fight until death.
    a) Goldfish b) Betta fish
    c) Angel fish d) Barbs
37. Match the pairs:

a)	Cyprinidae	i)	Botia striata
b)	Balitoridae	ii)	Ompak bimaculatus
c)	Cobitidae	iii)	Glyptothorax anamalaiensis
d)	Sisoridae	iv)	Salmostoma boopis
e)	Siluridae	v)	Xenentodon cancila
f)	Belonidae	vii)	Bhavania australis

g) Nandidae viii) Nandus nandus

h) Cichlidae viii) Etroplus suratensis

a) a=iv,b=vi,c=i,d=iii,e=ii,f=v,g=vii,h=vi

b) a=iv,b=iii,c=i,d=ii,e=v,f=viii,g=vii,h=vi

c) a=ii,b=vi,c=i,d=iv,e=iii,f=v,g=vii,h=viii

d) a=iv,b=ii,c=v,d=iii,e=i,f=vi,g=vii,h=viii

38. Members of family clupeidae are known as ....

a) Herrings b) Dipnoi

c) Minnows d) Featherbacks

39. In which family carps, minnows and barbs are included?

a) Family Sisoridae b) Family Cichlidae

c) Family Cyprinidae d) Family Belonidae

40. Global ornamental industry is dominated by ....

a) Marine water fishes b) Freshwater fishes

c) Brackish water fishes d) Euryhaline fishes

41. .... share 15% of the global ornamental fish trade.

a) Guppy & Neon tetra b) Gold fish

c) Angel fish d) Betta & Etroplus

42. .... contribute 69% to the total global ornamental fish trade.

a) Marine water ornamental fishes

b) Brackish water ornamental fishes

c) Freshwater ornamental fishes

d) Cord fishes

43. Tetras and Piranahas belongs to family ....

a) Siluridae b) Cyprinidae

c) Ambassidae d) Characidae

44. In aquarium plants are used for ....

a) Decoration b) Shade and shelter

c) Food d) All of these

45. Leaves of .... serve as food for ornamental fishes.

a) Pistia b) Hydrilla

c) Vallisneria d) All of these

46. Temperature range suitable for aquarium plants is ....
    a) 8-10 °C   b) 21-25 °C
    c) 7-8 °C   d) 23-40 °C
47. .... grow in cold as well as warm regions.
    a) Cryptocoryne ciliata   b) Cryptocoryne spiralis
    c) Aponogeton crispus   d) All of these
48. Match the pairs:
    a) Semi-terrestrial plant   i) Vallisneria
    b) Marginal plant   ii) Pistia
    c) Submerged plant   iii) Nymphaea
    d) Floating plant   iv) Typha
    e) Emergent plant   v) Ipomea
    a) a=v,b=iv,c=iii,d=ii,e=i   b) a=i,b=iii,c=iv,d-ii,e=v
    c) a=iv,b=v,c=i,d=ii,e=iii   d) a=iv,b=i,c=v,d=iii,e=ii
49. Which of the following is a floating plant ?
    a) Lemna   b) Ceratopteris
    c) Nymphaea   d) Both a & b
50. .... are cultivated by cuttings.
    a) Bacopa   b) Cabomba
    c) Both a & b   d) Pistia
51. Match the pairs:
    a) Acorus calamus   i) Amazon sword plant
    b) Aponogeton fenestralis   ii) Pigmy-chain sword
    c) Echinodorus paniculatus   iii) Hair grass
    d) Echinodorus tenellus   iv) Sweet flag
    e) Eleocharis acicularis   v) Madagascar leaf plant
    f) Sagittaria subulate   vi) Arrow head
    a) a=iii,b=i,c=v,d=ii,e=iv,f=vi   b) a=iv,b=v,c=i,d=ii,e=iii,f=vi
    c) a=vi,b=v,c=i,d=ii,e=iii,f=iv   d) a=iv,b=v,c=ii,d=i,e=iii,f=vi
52. Match the pairs:
    a) Acanthaceae   i) Vallisneria spiralis
    b) Alismaceae   ii) Eleocharis acicularis
    c) Araceae   iii) Hygrophila polysperma

d) Hydrocharitaceae iv) Sagittaria subulate
e) Cyperaceae v) Cabomba caroliniana
f) Nymphaeaceae vi) Acorus calamus
a) a=i,b=v,c=iii,d=vi,e=ii,f=iv b) a=iv,b=i,c=iii,d=vi,e=v,f=ii
c) a=i,b=v,c=iv,d=iii,e=ii,f=vi d) a=iii,b=iv,c=vi,d=i,e=ii,f=v

53. .... is a very fragile having no roots.
a) Ceratophyllum demersum b) Vallisneria spiralis
c) Sagittaria subulate d) All of these

54. Which of the following is known as 'Tape grass?'
a) Hygrophila polysperma b) Vallisneria spiralis
c) Ottelia alismoides d) Cabomba caroliniana

55. Match the pairs:
a) Vesicularia dubyana i) Water sprite
b) Azolla caroliniana ii) Crystal wort
c) Eichhornia crassipes iii) Java moss
d) Riccia fluitans iv) Water hyacinth
e) Nuphar luteum v) Yellow water lily
f) Ceratopteris thalictroides vi) Fairy moss
a) a=iii,b=ii,c=i,d=iv,e=vi,f=v b) a=vi,b=iii,c=ii,d=i,e=iv,f=v
c) a=iii,b=vi,c=iv,d=ii,e=v,f=i d) a=ii,b=iii,c=i,d=v,e=iv,f=vi

56. .... is known as 'Banana plant'.
a) Nymphoides aquatica b) Nuphar luteum
c) Nymphaea alba d) Lemna minor

57. To maintain the water temperature at the required level .... is used.
a) Thermostat b) Thermometer
c) Sand d) Electric bulb

58. .... is used to feed tubifex worms and chironomous larvae to the aquarium fishes.
a) Bog wood b) Peats
c) Feeding cup d) Rocks

59. .... is used to receive debris or uneaten food from the bottom of aquarium.
a) Siphon tube b) Dip tube
c) Scraper d) All of these

60. .... is required to give natural look, for appropriate rooting of natural plants.

a) Plastic plants b) Bedding material
c) Pebbles d) Marble chips

61. Aquarium aerator is connected to ....

a) Air stone b) Filtration unit
c) Toy d) One of them

62. .... serve to break the air down into a constant stream of smaller air bubbles.

a) Air stones b) Scraper
c) Siphon tube d) Dip tube

63. Piston pump is used to ....

a) Filter the water
b) To maintain water temperature
c) To supply air to tank
d) All of these

64. Ammonia concentration .... mg/l is toxic to fishes.

a) Less than 0.5 b) More than 0.5
c) Between 0.2 to 0.3 d) Less than 0.2

65. Filtration in aquarium is useful for ....

a) Maintenance of good quality of water
b) Partial correction when the water becomes toxic
c) Both a & b
d) To increase total dissolved solids

66. .... is operated by air pumps.

a) Box filter b) Foam filter
c) Both a & b d) None of these

67. Biological filtration in aquarium is done by ....

a) Fungi b) Algae
c) Bacteria d) Plankton

68. .... is used to extract ammonia from aquarium water.

a) Sand b) Zeolite
c) Gravel bed d) Air pump

69. Ozonizer is used in aquarium because ....
   a) It kills bacteria
   b) It increases the speed of nitrite break down
   c) Both a & b
   d) It maintains water hardness

70. In aquarium lighting is important because ....
   a) It is a key factor for colouration and maturation of fishes
   b) It is the basic energy source for plants
   c) Fishes need light to see their way and find food in aquarium
   d) All of these

71. Excess hardness of water prevents ....
   a) Development of fertilized eggs   b) Fish breeding
   c) Fish feeding   d) All of these

72. pH range suitable for freshwater aquarium is ....
   a) 5-6   b) 4-6
   c) 7-8   d) 9-10.5

73. .... is known as temporary hardness.
   a) Non-carbonate hardness   b) Carbonate hardness
   c) Total hardness   d) None of these

74. Temporary hardness is due to the presence of ....
   a) Bicarbonates and carbonates of calcium and magnesium salts in the water
   b) Bicarbonates of calcium and magnesium salts in the water
   c) Carbonates of calcium and magnesium salts in the water
   d) None of these

75. .... is the permanent hardness.
   a) Carbonate hardness   b) Non-carbonate hardness
   c) Total hardness   d) Chlorides and hardness

76. Hardness of water is due to ....
   a) Carbonates   b) Bicarbonates
   c) Chlorides   d) All of these

77. Most of the tropical aquarium fishes thrive well at ....
    a) 15-18°C b) 17-20 °C
    c) 22-30 °C d) 18-40 °C
78. At low temperature ....
    a) The fishes lose their appetite b) Immunity is reduced
    c) Both a & b d) Fishes breed
79. Accumulation of excess organic matter in aquarium causes ....
    a) Transparency b) Turbidity
    c) Decrease in salinity d) All of these
80. .... protect their eggs and young ones.
    a) Cichlids b) Anabantic
    c) Carps d) Both a & b
81. Which of the following is cannibalistic feeder?
    a) Barbs b) Tetras
    c) Both a & b d) Cyprinids
82. .... has adhesive eggs.
    a) Tetras b) Barbs
    c) Koicarp d) All of these
83. Which of the following is egg scatterer having non-adhesive eggs.
    a) Zebra danio b) Goldfishes
    c) Barbs d) Koi carp
84. .... are egg depositors.
    a) Rasbora b) Corydorus sp
    c) Cichlids d) All of these
85. Which of the following is a bubble nest builder?
    a) Gouramis b) Cichlids
    c) Koi carp d) Both a & b
86. .... gives birth to young ones.
    a) Egg layers b) Live bearers
    c) Cichlids d) Featherbacks
87. .... are categorized as 'easy to breed' type.
    a) Egg layers b) Live bearers
    c) Cyprinids d) Barbs

88. Example of livebearer fish is ....

a) Guppy | b) Platies
c) Mollies | d) All of these

89. Swordtails are ....

a) Livebearers | b) Egg layers
c) Both a & b | d) None of these

90. Breeding period of egg layers is ....

a) During June to August | b) December to March
c) Both a & b | d) Throughout the year

91. Breeding season of Anabantid is ....

a) During monsoon season | b) Throughout the year
c) During February to May | d) April to May

92. The Live bearers breed ....

a) Monsoon season | b) Summer season
c) Winter season | d) Throughout the year

93. Optimum water temperature for live bearer breeding is ....

a) 15-17 °C | b) 30-35 °C
c) 26-28 °C | d) 5-10 °C

94. 'Characin hook' is present on anal fin of ....

a) Males of tetras | b) Females of tetras
c) Males of Guppy fish | d) Females of Guppy fish

95. .... are egg scatterers.

a) Tetras | b) Barbs
c) Goldfishes | d) All of these

96. Which of the following eat own young ones and eggs?

a) Tetras | b) Goldfishes
c) Tilapias | d) Both a & b

97. Which plant is used in breeding of tetras?

a) Myriophyllum | b) Ceratophyllum
c) Water hyacinth | d) All of these

98. Maturity age of Goldfish is ....

a) First year | b) Second year
c) Third year | d) Fourth year

99. In Goldfish sex can be differentiated on the basis of ....
    a) Belly line b) Genital orifice
    c) Tubercles d) All of these
100. Fecundity of Goldfish is ....
    a) 50,000-60,000 eggs b) 2500-3000 eggs
    c) 500-1000 eggs d) 15000-20,000 eggs
101. Parental care is absent in ....
    a) Tilapia b) Zebra danio
    c) Betta fish d) Both a & b
102. In Corydorus ....
    a) Female is larger than male
    b) During breeding time the belly of female looks red
    c) Female has swollen belly
    d) All of these
103. Which of the following is representative of Cichlid?
    a) Cichlasoma meeki b) Symphysodon discus
    c) Apistogramaa ramirezi d) All of these
104. Angelfish is ....
    a) Piscivorous b) Herbivorous
    c) Planktivorous d) Carnivorous
105. Fecundity of Angelfish is ....
    a) 1000-15000 eggs at a time b) 500-700 eggs at a time
    c) 100-2000 eggs at a time d) 50-70 eggs at a time
106. For aquarium fish breeding tank is sterilized with ....
    a) Lime solution b) Dilute Sodium chloride
    c) Formalin d) Potassium permanganate
107. Scientific name of Red discus is ....
    a) Symphysodon aequifasciata b) Symphysodon discus
    c) Symphysodon rabitus d) None of these
108. Which of the following is mouth brooder?
    a) Haplochromis b) Tilapia
    c) Angelfish d) Both a & b

109. Female of which fish keep eggs in mouth?

a) Angel fish | b) Betta fish
c) Tilapia | d) Barbus

110. The breeding requirement of Aplocheilus is ....

a) Live foods | b) Floating plants
c) Sand | d) Both a & b

111. Which fishes are known as 'bubble nest builder'?

a) Carps | b) Labyrinth fishes
c) Featherbacks | d) All of these

112. .... is characterized by the presence of a labyrinthine chamber above the gills.

a) Family cichlidae | b) Family Siluridae
c) Family Anabantidae | d) Family Cyprinodontidae

113. Which of the following is known as 'Siamese fighting fish'?

a) Aplocheilus punchax | b) Rasbora daniconius
c) Angel fish | d) Betta splendens

114. Betta fish ....

a) Livebearer | b) Egg layer
c) Nest builder | d) Both b & c

115. Male of .... are aggressive in nature and tend to kill a female, if she is not ready to breed.

a) Blue Gourami | b) Tilapia
c) Angelfish | d) Betta fish

116. Scientifc name of 'Blue Gourami' is ....

a) Trichogaster tricopterus | b) Aplocheilus punchax
c) Symphysodon discus | d) Corydorus puleatus

117. Babies of .... feeds on the parent's slime.

a) Tilapia | b) Koi carp
c) Discus | d) Betta fish

118. Live bearers belongs to which family?

a) Hemiramphidae | b) Goodeidae
c) Poeciliidae | d) All of these

119. Match the pairs:

a) Goodeidae	i) Moolinisia sp.
b) Mollies	ii) Xiphophorus helleri
c) Platies	iii) Characodon sp
d) Swardtails	iv) Heterandria formosa
e) Mosquito fish	v) Xiphophorus maculatus
f) Anablepidae	vi) Anableps sp

a) a=iii,b=v,c=ii,d=iv,e=i,f-=vi
b) a=i,b=iii,c=iv,d=iii,e=vi,f=v
c) a=ii,b=i,c=iii,d=v,e=iv,f=vi
d) a=iii,b=i,c=v,d=ii,e=iv,f=vi

120. Guppies sexually mature at ....
a) Ten months
b) Three months
c) One month
d) Seven months

121. Life span of Guppy is ....
a) 6 years
b) 4.5 years
c) 3.5 years
d) 8.5 years

122. During transportation which absorbent is added in the medium?
a) Permutit
b) Synthetic amberlite
c) Clinoptilolite
d) All of these

123. Hermaphroditism is observed in ....
a) Clownfish
b) Parrot fish
c) Angel fish
d) All of these

124. In .... male and female sex organs are present at the same time.
a) Simultaneous hermaphroditism
b) Sequential hermaphroditism
c) Protandry
d) Protogyny

125. In .... the young ones are born in one sex, but can change into the opposite sex later.
a) Sequential hermaphroditism
b) Protandry
c) Protogyny
d) Both a & b

126. Similtaneous hermaphroditism is also known as ....
a) Sequential hermaphroditism
b) Synchronous hermaphroditism
c) Protogyny
d) Both a & b

127. Similtaneous hermaphroditism is observed in ....
   a) Hamlet fish
   b) Mangrove killfish
   c) Both a & b
   d) Parrot fish
128. Clown fish, Parrot fish and Angel fish shows ....
   a) Sequential hermaphroditism
   b) Simultaneous hermaphroditism
   c) Both a &b
   d) None of these
129. In .... the fish is born as male and then it changes into a female.
   a) Protogyny
   b) Protandry
   c) Protosyni
   d) None of these
130. Example of protandry is ....
   a) Grouper
   b) Gold fish
   c) Clown fish
   d) Parrot fish
131. Which of the following shows protogyny?
   a) Angel fish
   b) Grouper
   c) Parrot fish
   d) All of these
132. Example of bidirectional hermaphroditism is ....
   a) Grouper
   b) Angel fish
   c) Blue-banded goby
   d) Goldfish
133. Scientific name of 'Blue-banded goby' is ....
   a) Lythrypnus dalli
   b) Micrometrus aurora
   c) Micrometrus minimus
   d) Both b & c
134. In family Poecillidae anal fin is modified into ....
   a) Amplexes
   b) Gonopodium
   c) Clasper
   d) Anus
135. Which of the following is egg layer?
   a) Goldfish
   b) Tilapia
   c) Zebra fish
   d) All of these
136. Example of livebearer is ....
   a) Swordtail
   b) Guppies
   c) Platys
   d) All of these

137. Guppies and Mollies are ....

a) Ovoviviparous b) Viviparous

c) Egg layers d) Both a & b

138. .... fishes keep the embryo inside the body and supply nutrition through placenta.

a) Oviparous b) Viviparous

c) Both a & b d) Egg layers

139. The recommended dietary protein for ornamental fishes is ....

a) 10-12% b) 15-17%

c) 24-40% d) 65-70%

140. Example of live food organism is ....

a) Daphnia b) Artemia

c) Brachionus d) All of these

141. .... act as food for zooplankton and fishes.

a) Volvox b) Ulothrix

c) Cladophora d) All of these

142. .... are examples of microalgae

a) Chlorella b) Scendesmjus

c) Both a & b d) Moina

143. .... is a blue-green algae belonging to the cyanophyceae.

a) Spirulina b) Vallisneria

c) Daphnia d) Artemia

144. .... is rich source of protein, beta carote, vitamins and minerals.

a) Ulothrix b) Spirulina

c) Brachionus d) All of these

145. Spirulina contains ....

a) 62-71% protein b) 6-8% lipid

c) 15-17% carbohydrates d) All of these

146. .... may serve as a good source of food for herbivorous ornamental fishes.

a) Pistia b) Salvinia

c) Hydrilla d) All of these

147. Which of the following is popularly known as 'Brine shrimp'.
   a) Brachionus
   b) Spirulina
   c) Artemia
   d) Ulothrix

148. .... is a pelagic microorganism which serves as a delicious food for aquarium fishes.
   a) Fabrea salina
   b) Pila globose
   c) Potamogeton
   d) Vallisneria

149. Brachionus is ....
   a) Rotifer
   b) Mollusc
   c) Copepod
   d) Cladoceran

150. Which of the following is an example of Cladocera?
   a) Brachionus
   b) Paramecium
   c) Daphnia
   d) Mosquito larvae

151. *Moina micrura* belongs to which family?
   a) Copepodae
   b) Daphnidae
   c) Oligochaeta
   d) Moinidae

152. .... an example of crustacean zooplankton.
   a) Paramecium
   b) Tubifex
   c) Cyclops
   d) All of these

153. Match the pairs:

a) Gammus pulex	i) Oligochaeta
b) Calanus	ii) Blood worms
c) Alona sp	iii) Freshwater shrimp
d) Chironomous larvae	iv) Copepod
e) Anguillula	v) Cladoceran
f) Tubifex	vi) Microworm

   a) a=iii,b=iv,c=v,d=ii,e=vi,f=i
   b) a=iv,b=v,c=i,d=iii,e=ii,f=vi
   c) a=iii,b=iv,c=v,d=vi,e=ii,f=i
   d) a=iii,b=v,c=iv,d=vi,e=i,f=ii

154. Aquarium fish food is of which type?
   a) Dried
   b) Freeze dried
   c) Frozen
   d) All of these

155. Freeze-dried food include ....
   a) Daphina
   b) Tubifex worms
   c) Mosquito larvae
   d) All of these

156. Aquarium fish disease can be avoided by ....
   a) Avoiding overfeeding
   b) Avoiding over crowding
   c) Keeping fishes almost the same size
   d) All of these

157. Hospital tank is used to ....
   a) Keep the diseased fish
   b) Keep the dead fish
   c) Both a & b
   d) None of these

158. Symptom of diseased fish is ....
   a) Loss of appetite
   b) Abnormal swimming movement
   c) Change in colour
   d) All of these

159. Factor responsible for disease in aquarium fish is.......
   a) Nutritional disorders
   b) Physical injuries
   c) Effects due to environmental factors and pollutants
   d) All of these

160. Treatment for constipation in aquarium fish is ....
   a) To avoid overfeeding
   b) Bath for 4 hours in Epsom salt solution
   c) Both a & b
   d) Bath in formalin solution

161. .... is a side effect or due to oxygen super saturation.
   a) Gas bubble disease
   b) Bronchitis
   c) Acidosis
   d) Both b & c

162. .... is caused by haemorrhage in the capillaries of the eye sockets.
   a) Gas bubble disease
   b) Exophthalmia
   c) Fin rot
   d) Columnaris

163. Treatment of exophthalmia is ....
   a) A common salt bath to fish
   b) Bathing the eyes carefully with a 5% solution of Argyrol
   c) Both a & b
   d) Bathing in methylene blue solution

164. Finrot is caused by ....
   a) Fungi b) Protozoans
   c) Bacteria d) Worms

165. Tretment for fin rot is ....
   a) Swarb with 3% H2O2
   b) Treat with Acriflavin solution @ 0.6 gm/10 liters of water
   c) Combination of chloromycetin and penicillin
   d) All of these

166. Columnaris is also known as ....
   a) Cotton mouth disease b) White spot disease
   c) Ich disease d) Whirling disease

167. The best treatment for cloumnaris disease is ....
   a) To treat fish with common salt solution
   b) Swabbing the mouth with hydrogen peroxide
   c) Swabbing the mouth with iodine
   d) All of these

168. Septicaemia is a ....
   a) Blood poisoning b) Finrot disease
   c) Gas bubbling symptom d) Whirling disease

169. Tretament used for septicaemia is ....
   a) To treat fish with Sali solution b) To change the dirty water
   c) To treat the fish with antibiotics d) All of these

170. Dropsy in fishes is caused due to ....
   a) Malfunctioning of liver b) Malfunctioning of kidney
   c) Both a & b d) None of these

171. .... is responsible for causing dropsy in ornamental fishes.
   a) Aeromonas b) Costia necatrix
   c) Anchor worm d) Fish leech

172. A swollen abdomen is the characteristic sign of ....
    a) Dropsy disease
    b) White Spot Disease
    c) Fin rot disease
    d) Protozoan disease

173. The best treatment for dropsy disease is ....
    a) To disinfect the fish with streptomycin
    b) To disinfect the fish with chloromycetin
    c) Both a & b
    d) To keep the fish in formalin solution

174. .... is caused by overfeeding, overcrowding and feeding fishes with imbalanced diet.
    a) Ich disease
    b) Fish pox
    c) Tuberculosis
    d) Both a & b

175. Fish tuberculosis was first recognized in year ....
    a) 1905
    b) 1820
    c) 1853
    d) 1897

176. Symptom of viral haemorrhagic septicemia is.........
    a) Necrosis of the renal tubules
    b) The liver shows fat degeneration
    c) Kidney and muscles show edema
    d) All of these

177. The well known fungi of aquarium fishes are ....
    a) Achyla
    b) Saprolegina
    c) Both a & b
    c) Pseudomonas

178. Symptom of saprolegniasis is ....
    a) Fish become lethargic
    b) Fish become listless
    c) Fish become less responsive to external stimuli
    d) All of these

179. Treatment for saprolegniasis is ....
    a) Fish bath in 2% common salt solution
    b) Fish bath for 10-20 minutes in Potassium permanganate
    c) Fish bath in malachite green
    d) All of these

180. Ichthyosporidium is a ....
   a) Bacterium
   b) Fungus
   c) Virus
   d) Protozoan
181. Effective treatment for Ichthyosporidiasis is ....
   a) Use of phenoxy ethanol @1 ml/4 Lites of hot water
   b) Fish bath in quinine hydrochloride
   c) Fish bath in picric acid solution
   d) Use of lime solution in fish tank
182. Gillrot disease in fishes is also known as ....
   a) Branchiomycosis
   b) Whit spot disease
   c) Ich disease
   d) Costiasis
183. Velvet disease is caused by ....
   a) Oodinium
   b) Costia
   c) Fish leech
   d) Pseudomonas
184. Which treatment is effective for control of 'Velvet disease'?
   a) To dip fish for a second in 0.75 % malachite green solution
   b) To keep the fish in common salt solution
   c) To keep the fish in picric acid solution
   d) Both a & b
185. *Ichthyophthirius multifilis* is responsible for which disease?
   a) Ich disease
   b) Costiasis
   c) Dropsy
   d) Eye disease
186. Ich disease is also known as ....
   a) Ichthyophthiriasis
   b) Costiasis
   c) Whirling disease
   d) Fin rot disease
187. Treatment of Ichthyophthiriasis includes ....
   a) 3-4 drops of 2% formalin solution
   b) Treatment with quinine hydrochloride
   c) Treatment with string salt solution
   d) Both b & c
188. Which of the following is a protozoan disease?
   a) Costiasis
   b) Ichthyophthiriasis
   c) Both a & b
   d) Ulcer disease

189. Octomitus is a flagellate which lives in ....
    a) Fish scales
    b) Internal organs of fish
    c) Tail region of fish
    d) Fish fins

190. 'Whirling disease' is caused by ....
    a) *Costia necatrix*
    b) *Ichthyophthirius multifilis*
    c) *Myxosoma cerebralis*
    d) *Pseudomonas punctata*

191. 'Neon-disease' is caused by ....
    a) Plistophora hyphessobryconis
    b) Cyclochaeta sp.
    c) *Ichthyophthirius multifilis*
    d) *Saprolegina parasitica*

192. 'Shimmies' or 'sliminess' is caused by ....
    a) Costia
    b) Chilodon
    c) Cyclochaeta
    d) All of these

193. Which of the following is a monogenetic trematode?
    a) Clinostomum
    b) Diplostomum
    c) Dactylogyrus
    d) Costia necatrix

194. What is HACCP?
    a) A preventive system for assuring the safe production of food products
    b) Common-sense application of technical and scientific principles
    c) Both a & b
    d) None of these

195. Principles of HACCP are ....
    a) To conduct a hazard analysis
    b) To establish critical limits
    c) To identify the critical control points
    d) All of these

196. ....can not survive salt concentrations and do not grow in concentrations greater than 6%
    a) Halophilic bacteria
    b) Halotoleratnt bacteria
    c) Halophobic bacteria
    d) Both b & c

197. .... are salt loving bacteria that need salt greater than 2% concentration.
    a) Halophilic bacteria
    b) Halotolerant bacteria
    c) Halophobic bacteria
    d) Pink baccteria

198. Assesment of fish spoilage is done with help of ....
   a) Physical method
   b) Subjective method
   c) Organoleptic test
   d) All of these
199. Which of the following is a parasitic digenetic trematode?
   a) Clinostomum
   b) Diplostomum
   c) Dactylogyrus
   d) Both a & b
200. .... is a nematode fish parasite which infects live bearers?
   a) Clinostomium
   b) Camallanus
   c) Sanguinicola
   d) Both a & b
201. Which of the following is known as ' fish louse'?
   a) Lernaea
   b) Costia
   c) Argulus
   d) Chilodon
202. Argulus is a ....
   a) Crustacean
   b) Ectoparasite
   c) Both a & b
   d) Endoparasite
203. Which chemical is used for treatment of Argulosis ?
   a) Sodium chloride
   b) Butex
   c) Potassium permanganate
   d) All of these
204. Which of the following is a copepod crustacean parasite?
   a) Lernaea
   b) Argulus
   c) Clinostomum
   d) Diplostomum
205. .... is commonly known as 'Anchor worm'?
   a) Asellus
   b) Ergasilus
   c) Lernaea
   d) Costia
206. Treatment for Lernaeosis includes.........
   a) Treatment with 0.1% potassium permanganate solution
   b) Sodium chloride bath treatment @2%
   c) Both a & b
   d) Use of 6% formaline solution
207. .... infects gills, buccal cavity, operculum and fins.
   a) Argulus
   b) Fish leech
   c) Gyrodactylus
   d) Ergasilus

208. ....occurs as hereditary conditions.

a) Lack of fins
b) Swim bladder dysfunction
c) Skeletal deformities
d) All of these

209. Which of the following is a marine aquarium primarily meant for invertebrates,corals and anemones?

a) Reef aquarium
b) Fish only aquarium
c) All glass aquaria
d) Tropical aquarium

210. Fish only with live rock aquarium (FOWLR) is a ....

a) Freshwater aquarium
b) Marine aquarium
c) Brackish water aquarium
d) Reef aquarium

211. Which of the following is a brackish water ornamental fish?

a) Pufferfish
b) Goby
c) Flatfish
d) All of these

212. Which of the following is suitable for cold water aquarium?

a) Goldfish
b) Zebra fish
c) Rosy barb
d) All of these

213. .... are fitted together edge to edge using silicone rubber adhesive?

a) All glass tanks
b) Metal framed glass tanks
c) Both a & b
d) Ecotype aquaria

214. Stocking density of aquarium fishes is based on ....

a) Fish length
b) Volume of water
c) Both a & b
d) Water temperature

215. The ideal stocking density in home aquarium is ....

a) 1 cm long fish per 1 Liter of water
b) 10-inch-long fish per 1 gallon of water
c) 1 cm long fish per 5 Liters of water
d) 3-inch-long fish per 15 Liters of water

216. ....is a craft of arranging aquatic plants, rocks and stones in an aesthetically pleasing manner within an aquarium.

a) Aquascaping
b) Aquaranching
c) Aquaplanting
d) All of these

217. The fishes selected for aquarium should be ....
   a) Non-aggressive b) Non-predatory
   c) Calm and peaceful d) All of these
218. In aquarium avoid fishes with ....
   a) Tornfins b) White spots
   c) Bulging eyes d) All of these
219. .... are spherical black plastic balls used as filtration media in aquarium filters.
   a) Zeolites b) Bioballs
   c) Chemosensors d) Coaltoy
220. .... are alumino silicate minerals.
   a) Bioballs b) Zeolites
   c) Coral reefs d) All of these
221. .... adsorb ammonia from aquarium water.
   a) Zeolites b) Charcoal
   c) Sponge filter d) Corner filter
222. Which of the following adsorb calcium from hardwater of aquarium?
   a) Charcoal b) Sponge filter
   c) Bioballs d) Zeolites
223. The difficulty with marine fish keeping is.........
   a) They are highly priced
   b) They need stable environment
   c) Both a & b
   d) None of these
224. What is 'Live rock'?
   a) A piece of coral that is broken off from coral reef
   b) A filtration device
   c) Live sand
   d) Algae
225. Benefits of Live rock is ....
   a) It is a fantastic biological filter
   b) It provides hiding and living spaces for fish
   c) It can be source of food for saltwater fish
   d) All of these

226. .... is a process that is performed when adding new marine life to an aquarium.
   a) Bioassay b) Acclimation
   c) Aquascaping d) Bioranching
227. Invertebrates used in aquarium are ....
   a) Snapping shrimps b) Lobsters
   c) Octopus d) All of these
228. .... are described as 'terrestrial flowers' of the sea?
   a) Anemones b) Echinoderms
   c) Molluscs d) Corals
229. .... is an accessory tank of marine aquarium in which the mechanical equipment are kept.
   a) Hospital tank b) Sump
   c) Trickle filter d) Reef tank
230. The first attempt of keeping marine fish alive was made by ....
   a) Mrs. Thynne b) Mrs.Rioss
   c) Mrs.Elizabeth d) Mrs.Margaret
231. The number of watt in the light lamp for aquarium is ....
   a) 100 b) 200
   c) 50 d) 300
232. Lightning in aquarium is essential for .... hours.
   a) 10 b) 6
   c) 12 d) 25
233. The aquarium glass is cleaned by a ....
   a) Cotton b) Steel wood
   c) Picric acid d) Cloth
234. The fish do not eat its own egg is ....
   a) Molly b) Platy
   c) Cichlid d) Guppy
235. One of the following inorganic form used by the aquarium plants as nutrients is ....
   a) Suplhate b) Nitrate
   c) Carbonate d) Phosphate

236. The male fish which has a brood pouch beneath the tail is ....
   a) Cichlid b) Sea horse
   c) Swordtail d) Koi carp

237. Name the male ornamental fish which carries the eggs in its mouth is.........
   a) Arius b) Platy
   c) Swordtail d) Guppy

238. The BOD of the water increases, DO level is ....
   a) Increased b) Maintained
   c) Decreased d) No reaction

239. The time required for breeding ornamental fish after injection of pituitary extract is.........
   a) 8 hours b) 16 hours
   c) 4 hours d) 18 hours

240. Which is the largest aquarium in the world?
   a) Chimelong ocean kingdom (China)
   b) The Georgia aquarium (U.S.)
   c) Moscow oceanarium (Russia)
   d) Shedd aquarium (U.S.)

241. Which is the India's largest public aquarium?
   a) Taraporewala aquarium (Mumbai)
   b) Marine World, Thrissur (Kerala)
   c) Chennai oceanarium (Chennai)
   d) Ratnagiri aquarium (U.S.)

242. Which ornamental fish is banned in India?
   a) Discus b) Tilapia
   c) Piranah d) All of these

243. .... being first among ornamental fish producing country?
   a) Singapore b) Japan
   c) Russia d) India

244. Which Indian state has highest ornamental fish production?
   a) Maharashtra b) Gujarat
   c) West Bengal d) Tamil Nadu

245. The major objectives of the pilot project 'Mission ornamental Fishes-2017' are ....
   a) To promote culture of preferred species with cluster-based approach
   b) To create employment opportunities for the rural and peri-urban population
   c) To use modern technologies and innovations to make ornamental fisheries a thriving activity
   d) All of these

246. How many states were identified under Pilot project 'Mission ornamental Fishes-2017'?
   a) 11 b) 4
   c) 5 d) 9

247. India exports most of its ornamental fish to ....
   a) Singapore b) Japan
   c) United States d) All of these

248. Ornamental Fisheries Training and Research Institute is located at.........
   a) Bangalore b) Udaipur
   c) Mumbai d) Kochi

249. India's share to global aquarium fish export is ....
   a) 0.4% b) 1.3%
   c) 4.2% d) 6.9%

250. The major marine ornamental fish exporting countries are ....
   a) Indonesia b) Philippines
   d) Sri Lanka d) All of these

Answer Keys

1	a)	Philip Henry Gosse
2	d)	Anne Thynne
3	b)	London zoo
4	a)	Year 1951
5	b)	Mumbai
6	c)	Martin Horowitz
7	d)	Both a & b
8	c)	Wild collection
9	a)	Live rock
10	d)	All of these

11	d)	All of these
12	b)	Bitumen
13	d)	All of these
14	b)	Silicone sealant
15	a)	Squeezing gun
16	d)	All of these
17	a)	Community tank
18	b)	Cyprinidae
19	d)	All of these
20	a)	Botia
21	d)	All of these
22	b)	Mastacembelidae
23	a)	Nandidae
24	b)	Dwarf chameleon
25	a)	Tetradontidae
26	c)	Snakeheads
27	a)	Ambassidae
28	c)	Ambassidae
29	c)	Both a & b
30	a)	Poecilia reticulata
31	b)	Guppy
32	d)	Guppy
33	a)	Poecilidae
34	c)	Astronotus ocellatus
35	a)	Osphronemidae
36	b)	Betta fish
37	a)	a=iv,b=vi,c=i,d=iii,e=ii,f=v,g=vii,h=vi
38	a)	Herrings
39	c)	Family Cyprinidae
40	b)	Freshwater fishes
41	a)	Guppy & Neon tetra
42	c)	Freshwater ornamental fishes
43	d)	Characidae
44	d)	All of these
45	d)	All of these
46	b)	21-25 °C
47	d)	All of these
48	c)	a=iv,b=v,c=i,d=ii,e=iii
49	d)	Both a & b

50	c)	Both a & b
51	b)	a=iv,b=v,c=i,d=ii,e=iii,f=vi
52	d)	a=iii,b=iv,c=vi,d=i,e=ii,f=v
53	a)	Ceratophyllum demersum
54	b)	Vallisneria spiralis
55	c)	a=iii,b=vi,c=iv,d=ii,e=v,f=i
56	a)	Nymphoides aquatica
57	a)	Thermostat
58	c)	Feeding cup
59	b)	Dip tube
60	b)	Bedding material
61	d)	One of them
62	a)	Air stones
63	c)	To supply air to tank
64	b)	More than 0.5
65	c)	Both a & b
66	c)	Both a & b
67	c)	Bacteria
68	b)	Zeolite
69	c)	Both a & b
70	d)	All of these
71	a)	Development of fertilized eggs
72	c)	7-8
73	b)	Carbonate hardness
74	a)	Bicarbonates and carbonates of calcium and magnesium salts in the water
75	b)	Non-carbonate hardness
76	d)	All of these
77	c)	22-30 °C
78	c)	Both a & b
79	b)	Turbidity
80	d)	Both a & b
81	c)	Both a & b
82	d)	All of these
83	a)	Zebra danio
84	d)	All of these
85	a)	Gouramis
86	b)	Live bearers
87	b)	Live bearers
88	d)	All of these

89	a)	Livebearers
90	c)	Both a & b
91	a)	During monsoon season
92	d)	Throughout the year
93	c)	26-28 oC
94	a)	Males of tetras
95	d)	All of these
96	d)	Both a & b
97	a)	All of these
98	b)	Second year
99	d)	All of these
100	b)	2500-3000 eggs
101	b)	Zebra danio
102	d)	All of these
103	d)	All of these
104	d)	Carnivorous
105	c)	100-2000 eggs at a time
106	d)	Potassium permanganate
107	b)	Symphysodon discus
108	d)	Both a & b
109	c)	Tilapia
110	d)	Both a & b
111	b)	Labyrinth fishes
112	c)	Family Anabantidae
113	d)	Betta splendens
114	d)	Both b & c
115	a)	Blue Gourami
116	a)	Trichogaster tricopterus
117	c)	Discus
118	d)	All of these
119	d)	a=iii,b=i.c=v,d=ii,e=iv,f=vi
120	b)	Three months
121	c)	3.5 years
122	d)	All of these
123	d)	All of these
124	a)	Simultaneous hermaphroditism
125	a)	Sequential hermaphroditism
126	b)	Synchronous hermaphroditism
127	c)	Both a & b

128	a)	Sequential hermaphroditism
129	b)	Protandry
130	c)	Clown fish
131	d)	All of these
132	c)	Blue-banded goby
133	a)	Lythrypnus dalli
134	b)	Gonopodium
135	d)	All of these
136	d)	All of these
137	a)	Ovoviviparous
138	b)	Viviparous
139	c)	24-40%
140	d)	All of these
141	d)	All of these
142	c)	Both a & b
143	a)	Spirulina
144	b)	Spirulina
145	d)	All of these
146	d)	All of these
147	c)	Artemia
148	a)	Fabrea salina
149	a)	Rotifer
150	c)	Daphnia
151	b)	Daphnidae
152	c)	Cyclops
153	a)	a=iii,b=iv,c=v,d=ii,e=vi,f=i
154	d)	All of these
155	d)	All of these
156	d)	All of these
157	a)	Keep the diseased fish
158	d)	All of these
159	d)	All of these
160	c)	Both a & b
161	a)	Gas bubble disease
162	b)	Exophthalmia
163	c)	Both a & b
164	c)	Bacteria
165	d)	All of these
166	a)	Cotton mouth disease

167	d)	All of these
168	a)	Blood poisoning
169	d)	All of these
170	c)	Both a & b
171	a)	Aeromonas
172	a)	Dropsy disease
173	c)	Both a & b
174	c)	Tuberculosis
175	d)	1897
176	d)	All of these
177	c)	Both a & b
178	d)	All of these
179	d)	All of these
180	b)	Fungus
181	a)	Use of phenoxy ethanol @1 ml/4 Lites of hot water
182	a)	Branchiomycosis
183	a)	Oodinium
184	d)	Both a & b
185	a)	Ich disease
186	a)	Ichthyophthiriasis
187	d)	Both b & c
188	c)	Both a & b
189	b)	Internal organs of fish
190	c)	Myxosoma cerebralis
191	a)	Plistophora hyphessobryconis
192	d)	All of these
193	c)	Dactylogyrus
194	c)	Both a & b
195	d)	All of these
196	c)	Halophobic bacteria
197	a)	Halophilic bacteria
198	d)	All of these
199	d)	Both a & b
200	b)	Camallanus
201	c)	Argulus
202	c)	Both a & b
203	d)	All of these
204	a)	Lernaea
205	c)	Lernaea

206	c)	Both a & b
207	d)	Ergasilus
208	d)	All of these
209	a)	Reef aquarium
210	b)	Marine aquarium
211	d)	All of these
212	d)	All of these
213	a)	All glass tanks
214	c)	Both a & b
215	a)	1 cm long fish per 1 Liter of water
216	a)	Aquascaping
217	d)	All of these
218	d)	All of these
219	b)	Bioballs
220	b)	Zeolites
221	a)	Zeolites
222	d)	Zeolites
223	c)	Both a & b
224	a)	A piece of coral that is broken off from coral reef
225	d)	All of these
226	b)	Acclimation
227	d)	All of these
228	a)	Anemones
229	b)	Sump
230	a)	Mrs. Thynne
231	c)	50
232	b)	6
233	a)	Cotton
234	c)	Cichlid
235	b)	Nitrate
236	b)	Sea horse
237	a)	Arius
238	c)	Decreased
239	c)	4 hours
240	a)	Chimelong ocean kingdom (Chin
241	b)	Marine World, Thrissur (Keral
242	c)	Piranah
243	a)	Singapore
244	c)	West Bengal

245	d)	All of these
246	d)	9
247	d)	All of these
248	b)	Udaipur
249	a)	0.4%
250	d)	All of these

Table 1: Commercially important freshwater ornamental fishes of India

Sr.No.	Scientific Name	Common Name
1	*Barbus arulius*	Aruli barb
2	*Barbus oligolepis*	Checker barb
3	*Barbus tetrazona*	Tiger barb
4	*Barbus cummingi*	Cumming's barb
5	*Barbus conchonius*	Rosy barb
6	*Barbus chola*	Chola barb
7	*Puntius titteye*	Cherry barb
8	*Danio malabaricus*	Pearl danio
9	*Danio albolineata*	Zebra fish
10	*Brachydanio rerio*	Zebra fish
11	*Colisa lalia*	Dwarf Gourami
12	*Colisa chuna*	Honey Gourami
13	*Colisa fasciata*	Indian Gourami
14	*Colisa labiosa*	Thick-lipped Gourami
15	*Trichopodus trichopterus*	Three spot Gourami
16	*Trichopodus leeri*	Pearl Gourami
17	*Trichipodus microlepis*	Moonlight Gourami
18	*Heleostoma temmincki*	Kissing Gourami
19	*Macrognathus aculeatus*	Pearl cock eel/Spotted spiny eel
20	*Mastacembelus armatus*	Zigzag eel
21	*Monopterus travancoricus*	Malabar pufferfish
22	*Poecilia reticulata*	Guppy
23	*Poecilia sphenops*	Black molly
24	*Poecilia latipinna*	Sail fin molly
25	*Poecilia velifera*	Lyre tail molly
26	*Carassius auratus*	Goldfish
27	*Xiphophorus maculatus*	Southern platy
28	*Xiphophorus xiphidium*	Swordtail platy
29	*Xiphophorus variatus*	Variatus platy
30	*Xiphophorus helleri*	Swordtails

31	*Astronotus ocellatus*	Oscar
32	*Betta splendens*	Siamese fighting fish
33	*Heros efasciatus*	Severum
34	*Pterophyllum scalare*	Angelfish
35	*Symphysodon discus*	Heckel discus
36	*Symphysodon aequifasciata*	Green discus
37	*Paracheirodon innesi*	Neon tetras
38	*Paracheirodon axelrodi*	Cardinal tetra
39	*Hyphessobrycon herbetaxelrodi*	Black neon
40	*Aphyocharax anisitsi*	Blood fin
41	*Cheirodon erythrozonus*	Glow light tetra
42	*Hyphessobrycon pulchripinnis*	Lemon tetra
43	*Hyphessobrycon callistus*	Serpae tetra
44	*Gymnocorymbus ternetzi*	Black tetra
45	*Rasbora daniconeus*	Slender Rasbora
46	*Rasbora heteromorpha*	Red Rasbora
47	*Rasbora pauciperforata*	Glowlight Rasbora
48	*Rasbora dorsicellata*	Eye spot Rasbora
49	*Rasbora elegans*	Yellow Rasbora
50	*Rasbora maculate*	Spotted Rasbora
51	*Rasbora trilineata*	Scissortail Rasbora
52	*Esomus danricus*	Flying Barb
53	*Tanichthys albonubes*	White cloud mountain minnow
54	*Labeo chrysophekadion*	Black shark
55	*Labeo bicolor*	Red tailed Black Shark
56	*Labeo erythrurus*	Red Finned Shark
57	*Rhodeus amarus*	Bitterling
58	*Carassius auratus*	Goldfish
59	*Gymncorymbus ternetzi*	Widow tetra
60	*Hyphessobrycon callistus*	Serpae tetra
61	*Hyphessobrycon pulchripinnis*	Lemon tetra
62	*Hyphessobrycon flammeus*	Flame-tetra
63	*Hyphessobrycon scholzei*	Black Line Tetra
64	*Paracheirodon innesi*	Neon tetra
65	*Cheirodon axelrodi*	Cardinal tetra
66	*Pristella riddle*	x-ray fish
67	*Hemigrammus pulcher*	Pretty tetra
68	*Hemigrammus caudovittatus*	Buenos aires tetra
69	*Megalamphodus megalopterus*	Black phantom tetra

Table 2: Commercially important marine ornamental fishes of India

Sr.No.	Scientific Name	Common Name
1	*Amblyeleotris gymnocephala*	Masked shrimp goby
2	*Mahidolia mystacina*	Flagfin prawn goby
3	*Paragobiodon echinocephalus*	Redhead goby
4	*Periophthalmus argentilineatus*	Barred mud skipper
5	*Priolepis eugenius*	Noble goby
6	*Priolepis inhaca*	Brick goby
7	*Trimma annosum*	Grey beared pygmy goby
8	*Valenciennea muralis*	Mural gobi
9	*Valenciennea sexguttata*	Sixspot goby
10	*Valenciennea strigata*	Blueband goby
11	*Balistapus undulatus*	Orange-lined triggerfish
12	*Balistes vetula*	Queen triggerfish
13	*Balistoides conspicillum*	Clown triggerfish
14	*Sufflamen chrysopterum*	Halfmoon trigger fish
15	*Sufflamen fraenatum*	Masked trigger fish
16	*Andamia reyi*	Suckerlip blenny
17	*Blenniellia periophthalmus*	Bluedashed rock skipper
18	*Cirripectes castaneus*	Chest nut eyelash blenny
19	*Cirripectes stigmaticus*	Red streaked blenny
20	*Ecsenius midas*	Persian blenny
21	*Salarias fasciatus*	Jewelled blenny
22	*Centropyge bicolor*	Bicolor angelfish
23	*Centropyge eibli*	Blacktain angelfish
24	*Centropyge multispinis*	Dusky angelfish
25	*Pomacanthus annularis*	Blue ring angelfish
26	*Pomacanthus imperator*	Emperor angelfish
27	*Pomacanthus semicirculatus*	Semicircle angelfish
28	*Amphiprion ocellaris*	Clown anemone fish
29	*Amphiprion percula*	Orange clown fish
30	*Amphiprion bicinctus*	Two band anemone fish
31	*Amphiprion ephippium*	Saddle anemone fish

32	*Amphiprion frenatus*	Tomato clown fish
33	*Amphiprion nigripes*	Maldive anemone fish
34	*Amphiprion sebae*	Sebae anemone fish
35	*Dascyllus aruanus*	Damsel fish
36	*Pomacentrus moluccensis*	Lemon Damsel
37	*Chrysiptera parasema*	Yellow tail blue Damselfish
38	*Chrysiptera cyanea*	Blue Devil Damselfish
39	*Chromis viridis*	Blue/Green chromis
40	*Chaetodon capistratus*	Four eyed butterfly fish

4

Live Feed Culture

1. Live feed is also known as ....
   a) Natural food b) Fish food organisms
   c) Both a & b d) Algae
2. Live feed are organisms used as food by ....
   a) Prawns b) Carps
   c) Oysters d) All of these
3. The common phytoplankton used as live feed are............
   a) Eudorina b) Volvox
   c) Pediastrum d) All of these
4. The common zooplankton used as live feed are ....
   a) Vorticella b) Artemia
   c) Cyclops d) All of these
5. Aretmia is commonly known as ....
   a) Brine shrimp b) Sea monkey
   c) Both a & b d) Copepod
6. The life span of artemia is ....
   a) Six months b) Ten months
   c) One month d) Twelve months
7. .... was discovered in 1831 in the Great Salt Lake in the Utah Desert
   a) Cyclops b) Brachionus
   c) Artemia d) Chlorella
8. The natural population of artemia found in India is known as ....
   a) Artemia parthenogenetica b) Artemia parasitica
   c) Artemia vulgaris d) Artemia vulgaris

9. Artemia is cultured by which method?
   a) Culture in salterns
   b) Culture in raceway tanks
   c) Culture in cages
   d) Both a & b
10. Artemia can be cultured in .... by intensive culture method.
   a) Salt pans
   b) Cement tanks
   c) Bundhs
   d) Both a & b
11. In intensive culture artemia is fed with ....
   a) Rice bran
   b) Yeast
   c) Chlorella
   d) All of these
12. Daphnia is a ....
   a) Cladoceran crustacean
   b) Freshwater monkey
   c) Shrimp
   d) Algae
13. .... functions as feed organism for Indian major carps and prawns.
   a) Daphnia
   b) Forage fish
   c) Both a & b
   d) Weed fishes
14. .... is a freshwater zooplankton which reproduces by parthenogenesis.
   a) Spirulina
   b) Chlorella
   c) Daphnia
   d) All of these
15. The Daphnia feed on ....
   a) Chlorella
   b) Chlamydomonas
   c) Aretmia
   d) Both a & b
16. Which of the following is a rotifer?
   a) Cyclop
   b) Brachionus
   c) Chlorella
   d) Spirulina
17. .... is a copepod crustacean.
   a) Brachionus
   b) Daphnia
   c) Artemia
   d) Cyclops
18. .... is an oligochaete annelid.
   a) Artemia
   b) Tubifex
   c) Brachionus
   d) Both a & b
19. Spirulina is a ....
   a) Blue green algae
   b) Rotifer
   c) Cladoceran crustacean
   d) Branchiopod crustacean

20. Spirulina naturally found in ....
    a) Freshwaters b) Brackishwaters
    c) Marine waters d) All of these
21. Biomass of spirulina contains ....
    a) Proteins b) Vitamins
    c) Minerals d) All of these
22. .... is a spherical, non-motile unicellular green algae.
    a) Artemia b) Daphnia
    c) Brachionus d) Chlorella
23. Chlorella found in ....
    a) Marine water b) Freshwater
    c) Sewage water d) All of these
24. .... is cultured using Sorokin-krauss medium.
    a) Chlorella b) Daphnia
    c) Artemia d) Both a & b
25. .... is a coenobial green algae found in lentic waters.
    a) Daphnia b) Artemia
    c) Scenedesmus d) Both a & b
26. The terminal cells of .... have two extensions of cell wall called spines.
    a) Scendesmus b) Daphnia
    c) Chlorella d) Brachionus
27. Which of the following is an example of copepod?
    a) Cyclops b) Diaptomus
    c) Canthocamptus d) All of these
28. .... act as excellent diet for the culture of copepods?
    a) Algae b) Diatoms
    c) Artemia d) Both a & b
29. 'Stable tea', an excreta of horse is used to culture ....
    a) Cladocerans b) Artemia
    c) Chlorella d) All of these

30. Cladocerans are commonly known as ....
    a) Water fleas b) Sea monkey
    c) Rotatoria d) Both a & b
31. Tubifex belongs to which family?
    a) Vonantidae b) Naididae
    c) Belonidae d) None of these
32. Tubifex worms feed on ....
    a) Decaying organic matter b) Detritus
    c) Vegetable matter d) All of these
33. .... are commonly called as 'wheel animalcules'.
    a) Cladocerans b) Rotifers
    c) Copepods d) Tubifex
34. .... is the most known form of rotifer.
    a) Brachionus b) Spirulina
    c) Chlorella d) Diatom
35. Moina and Daphnia are ....
    a) Protozoans b) Rotifers
    c) Cladocerans d) None of these
36. .... is a insect belonging to order Diptera of class insecta.
    a) Tubifex b) Chironomid larva
    c) Brachionus d) Both a & b
37. .... make an ideal suited diet for brooders of ornamental fishes.
    a) Agar agar b) Costia
    c) Tubifex d) Gyrodactylus
38. .... refers to microscopic single celled animalcules belonging to the class ciliata of phylum protozoa.
    a) Rotifera b) Infusoria
    c) Ciliates d) All of these
39. .... are the most common forms of freshwater infusoria.
    a) Paramecium b) Stylonychia
    c) Fabrea d) Both a & b
40. Which of the following is an example of marine infusoria?
    a) Fabrea b) Euplotes
    c) Both a & b d) Stylonychia

41. Which of the following is a life stage of artemia?
    a) Cyst
    b) Nauplii
    c) Juvenile
    d) All of these
42. Artemia is ....
    a) Filter feeder
    b) Continuous feeder
    c) Both a & b
    d) None of these
43. Freshly hatched artemia nauplius has ....
    a) No mouth
    b) Can not feed
    c) Developed mouth
    d) Both a & b
44. Rotifers are important live feed in aquaculture because ....
    a) They are planktonic in nature
    b) Can tolerate to wide range of environmental conditions
    c) High reproducibility
    d) All of these
45. .... are chlorophyll bearing unicellular or multi-cellular plants.
    a) Algae
    b) Zooplankton
    c) Shrimps
    d) Both b & c
46. Which of the following is a multicellular alga?
    a) Giant kelp
    b) Brown algae
    c) Diatoms
    d) Both a &b
47. Unicellular examples of algae include ....
    a) Diatoms
    b) Euglenophyta
    c) Dinoflagellates
    d) All of these
48. Which of the following is unicellular algae?
    a) Chaetoceros
    b) Skeletonema
    c) Both a & b
    d) Daphnia
49. Match the pairs:

a) Rotifer	i) Euterpina
b) Infusoria	ii) Chlorella
c) Microalgae	iii) Chironomid larvae
d) Copepod	iv) Asplanchna
e) Cladoceran	v) Moina
f) Insect	vi) Paramecium

a) a=iv,b=vi,c=i,d=ii,e=iii,f=v
b) a=iii,b=vi,c=i,d=ii,e=iv,f=v
c) a=iv,b=vi,c=ii,d=i,e=v,f=iii
d) a=iv,b=vi,c=i,d=v,e=ii,f=iii

50. Example of micro algae is ....
a) Chlamydomonas
b) Isochrysis
c) Moina
d) Both a & b

51. Which of the following contains high protein?
a) Spirulina
b) Spirogyra
c) Schizochytrium
d) Pavlova sp.

52. Which of the following is a cladoceran?
a) Bosmia longirostris
b) Disphanosoma birgei
c) Daphnia lumholtzi
d) All of these

53. Life span of rotifers is ....
a) 5-12 days
b) 20-25 days
c) 2-3 days
d) 25-35 days

54. The nutritional content of moina depends on ....
a) Their size
b) Type of food they are receiving
c) Water temperature
d) Both a & b

55. The chironomid larvae contains ....
a) Protein
b) Lipid
c) Vitamins
d) All of these

56. The most important factor governing the nutritional quality of live feeds for aquaculture practices is ....
a) Essential fatty acid content
b) Life span of live feed
c) Size of live feed
d) Both a & b

57. Eicosapentanoic acid (EPa) and Docosahexanoic acid (DHa) are commonly known as ....
a) Highly unsaturated fatty acids (HUFa)
b) Proteins
c) Vitamins
d) None of these

58. Main suborders of copepods found in brackish water are ....
    a) Calanoids b) Cyclopoids
    c) Harpacticoids d) All of these
59. Life span of copepod i s.........
    a) 5-10 days b) 2-3 days
    c) 10-12 months d) 40-50 days
60. Major constraints in live feed culture are ....
    a) Difficulties in getting pure strain
    b) Lack of infrastructure facility like controlled environmental laboratory
    c) Live feed also acts as carrier of diseases to the fish larvae
    d) All of these

Answer Keys

1	c)	Both a & b
2	d)	All of these
3	d)	All of these
4	d)	All of these
5	c)	Both a & b
6	a)	Six months
7	c)	Artemia
8-	a)	Artemia parthenogenetica
9	d)	Both a & b
10	b)	Cement tanks
11	d)	All of these
12	a)	Cladoceran crustacean
13	a)	Daphnia
14	c)	Daphnia
15	d)	Both a & b
16	b)	Brachionus
17	d)	Cyclops
18	b)	Tubifex
19	a)	Blue green algae
20	d)	All of these
21	d)	All of these
22	d)	Chlorella
23	b)	Freshwater

24	a)	Chlorella
25	c)	Scenedesmus
26	a)	Scendesmus
27	d)	All of these
28	d)	Both a & b
29	a)	Cladocerans
30	a)	Water fleas
31	b)	Naididae
32	d)	All of these
33	b)	Rotifers
34	a)	Brachionus
35	c)	Cladocerans
36	b)	Chironomid larva
37	c)	Tubifex
38	b)	Infusoria
39	d)	Both a & b
40	c)	Both a & b
41	d)	All of these
42	c)	Both a & b
43	d)	Both a & b
44	d)	All of these
45	a)	Algae
46	d)	Both a &b
47	d)	All of these
48	c)	Both a & b
49	c)	a=iv,b=vi,c=ii,d=i,e=v,f=iii
50	d)	Both a & b
51	a)	Spirulina
52	d)	All of these
53	a)	5-12 days
54	d)	Both a & b
55	d)	All of these
56	a)	Essential fatty acid content
57	a)	Highly unsaturated fatty acids (HUFa)
58	d)	All of these
59	d)	40-50 days
60	d)	All of these

5

Pearl Culture

1. Which of the following is an example of freshwater mussel?
   a) Lamellidens marginalis
   b) Lamellidens corrianus
   c) Pictada fucata
   d) Both a & b
2. .... is secreted by the mantle of the pearl oyster as a protection against foreign objects.
   a) Pearl
   b) Swarm
   c) Prismatic layer
   d) Nacreous layer
3. Shell of pearl oyster is composed of .... layers.
   a) Two
   b) Three
   c) Four
   d) Five
4. Which is the outermost layer of shell?
   a) Prismatic layer
   b) Nacreous layer
   c) Periostracum layer
   d) None of these
5. .... is the outermost, greenish-brown, thin, translucent layer made up of conchiolin.
   a) Nacreous layer
   b) Mantle
   c) Periostracum
   d) Perismatic layer
6. Which is the middle layer of the shell secreted by mantle?
   a) Nacreous layer
   b) Prismatic layer
   c) Periostracum
   d) Both a & b
7. The innermost layer of shell of pearl is known as ....
   a) Nacreous layer
   b) Prismatic layer
   c) Periostracum
   d) Conchiolin
8. Which of the following layer is known as 'mother of pearl'?
   a) Prismatic layer
   b) Conchiolin
   c) Nacreous layer
   d) Periostracum

9. Which shell layer is responsible for pearl formation?
   a) Nacreous layer b) Prismatic layer
   c) Periostracum d) Both b & c
10. Size of pearl is directly proportional to ....
   a) Degree of irritation caused by foreign agents
   b) Water temperature
   c) Movement of pearl oyster
   d) Wave action
11. .... is a natural gem that is produced by a pearl oyster.
   a) Pearl b) Nacreours
   c) Raft d) Spat
12. The types of pearls include ....
   a) Akoya prearls b) Freshwater pearls
   c) South pearls d) All of these
13. Conditions required for high quality pearl production are ....
   a) A reliable source of pearl oysters b) A suitable location or site
   c) Access to grafting technicians d) All of these
14. The pearl is formed of ....
   a) Spat b) Nacre
   c) Both a & b d) None of these
15. The nacre is formed of ....
   a) Calcium carbonate b) Conchiolin
   c) Both a & b d) None of these
16. Which of the following is a marine pearl oyster?
   a) Pinctada vulgaris b) Pictada fucata
   c) Placuna placenta d) All of these
17. Pearl oysters are ....
   a) Bivalve b) Sedentary animals
   c) Both a & b d) Gastropods
18. Young oyster is known as ....
   a) Spats b) Nacre
   c) Lingha pearl d) Oriental pearl

19. Pearl culture involves ....
    a) Oyster collection
    b) Preparation of graft tissue
    c) Preparation of nucleus and implantation
    d) All of these
20. The piece of tissue which is inserted into the oyster is called ....
    a) Graft tissue b) Spats
    c) Nacre d) Mantle epithelium
21. .... is cut off from the mantle of another oyster.
    a) Graft tissue b) Nucleus
    c) Spats d) Swarm
22. .... is a foreign material which is inserted into the oyster.
    a) Nucleus b) Graft tissue
    c) Spat d) Brachiolaria
23. .... is referred to be the father of pearl industry in Japan.
    a) Kokichi Mikimoto b) Fritz Zernike
    c) Tukichi Nishikawa d) Luis Geffer
24. For the first time idea of pearl industry was evolved in ....
    a) Japan b) India
    c) China d) Russia
25. .... was the first person to get spherical artificial pearl.
    a) Kokichi Mikimoto b) Tokichi Nishikawa
    c) James Sherrington d) Paul Rossy
26. In India the pearl oyster are obtained from reefs in ....
    a) The Gulf of Mannar b) The Gulf of Kutch
    c) Pak Bay d) All of these
27. Pearl comprises.........
    a) Water b) Organic matter
    c) Calcium carbonate d) All of these
28. The best quality of pearls is ....
    a) Lingha pearl b) Seed pearls
    c) Blister pearls d) All of these

29. .... is obtained from marine oysters.

a) Nucleus b) Gobid mat

c) Lingha pearl d) None of these

30. .... has developed technology of freshwater pearl culture from common freshwater mussel.

a) Central Institute of Brackish water Aquaculture

b) Central Inland Fisheries Research Institute

c) Bay of Bengal-IGO

d) Central Institute of Freshwater Aquaculture

31. Natural enemies of pearl oysters are.........

a) Octopus b) Eel

c) Barnacles d) All of these

32. Shell holder is required to ....

a) Hold the mussels b) Insert nucleus

c) cut spatula d) Opening of shell

33. .... is used to remove the pearls.

a) Cell inserter b) Spatula

c) Nacre d) Spat

34. .... is used to keep the oyster open for the duration of operation.

a) Spatula b) Shell speculum

c) Retractor d) Graft lifting needle

35. .... is known as the gold lip or silver pearl oyster.

a) Pinctada maxima b) Pinctada margaritifera

c) Pinctada funca d) Pinctada vulgaris

36. .... is known as the black-lip oyster.

a) Pinctada margaritifera b) Pinctada maxima

c) Pinctada vulgaris d) Pteria penguin

37. The pearl oyster reefs in the Gulf of Kutch are known as ....

a) Pyster b) Bheries

c) Khaddas d) Klister

38. .... means a hard bottom of coral and rocky area with an admixture of mud and sand.

a) Glitter
b) Bundhs
c) Khadda
d) Both a & b

39. Factors responsible for the production of coloured pearls are ....

a) The depth of oyster culture and light penetration
b) Availability and quality of food phytoplankton content
c) Amount of trace element present
d) All of these

40. The colour of the pearl is related to ....

a) Metallic elements present in the oysters environment
b) pH of water
c) TDS and hardness of water
d) Turbidity of water

41. Use of pearl are ....

a) Used as an ornament and a symbol of grandeur
b) Used in ornamental handicraft
c) Used as medicine for surgery, paediatrics etc
d) All of these

42. In .... a project on pearl culture at Tuticorin Research Centra of CMFRI was initiated.

a) 1968
b) 1965
c) 1977
d) 1972

43. In India production of first cultured marine pearl was started in.........

a) 1970
b) 1973
c) 1976
d) 1979

44. .... is located vertically at the proximal end of the foot.

a) Byssal gland
b) Pineal gland
c) Racemose gland
d) Mushroom gland

45. Which of the following is a method of pearl oyster rearing?

a) Raft culture
b) Rack culture
c) Pen culture
d) Both a & b

46. Recently, scientist have succeeded in perfecting a simple technique for value added marine pearls called ....
    a) Mab-Bay Pearls   b) Von-Bayer Pearls
    c) Mint-Bay Pearls   d) Moxzaic-Bay Pearls
47. Genus Pinctada belongs to which family?
    a) Family Pteriidae   b) Family Crossotrceae
    c) Family Penaedeae   d) Family Cichilidae
48. Pearl banks in Gulf of Mannar are known as ....
    a) Paars   b) Khddas
    c) Striidis   d) Moans
49 .... are kept in post-operative care unit.
    a) Vultures   b) Implanted mussels
    c) Spats   d) Elver
50. In pearl mussel farming temperature of medium should be ....
    a) 7-8 ^{0}C   b) 25-30^{0}C
    c) 12-15^{0}C   d) 15-17^{0}C

Answer Keys

1	d)	Both a & b
2	a)	Pearl
3	b)	Three
4	c)	Periostracum layer
5	c)	Periostracum
6	b)	Prismatic layer
7	a)	Nacreous layer
8	c)	Nacreous layer
9	a)	Nacreous layer
10	a)	Degree of irritation caused by foreign agents
11	a)	Pearl
12	d)	All of these
13	d)	All of these
14	b)	Nacre
15	c)	Both a & b
16	d)	All of these
17	c)	Both a & b
18	a)	Spats

19	d)	All of these
20	a)	Graft tissue
21	a)	Graft tissue
22	a)	Nucleus
23	a)	Kokichi Mikimoto
24	a)	Japan
25	b)	Tokichi Nishikawa
26	d)	All of these
27	d)	All of these
28	a)	Lingha pearl
29	c)	Lingha pearl
30	d)	Central Institute of Freshwater Aquaculture
31	d)	All of these
32	a)	Hold the mussels
33	b)	Spatula
34	b)	Shell speculum
35	a)	Pinctada maxima
36	a)	Pinctada margaritifera
37	c)	Khaddas
38	c)	Khadda
39	d)	All of these
40	a)	Metallic elements present in the oysters environment
41	d)	All of these
42	d)	1972
43	b)	1973
44	a)	Byssal gland
45	d)	Both a & b
46	a)	Mab-Bay Pearls
47	a)	Family Pteriidae
48	a)	Paars
49	b)	Implanted mussels
50	b)	25-30^{0}C

6

Corals and Coral Reefs

1. Corals are ....

   a) Animals b) Plants

   c) Diatoms d) Plankton

2. Corals belongs to which phylum?

   a) Protozoa b) Echinodermata

   c) Cnidaria

3. The corals consists of a sac-like structure called ....

   a) Septa b) Polyp

   c) Haloclone d) Nephris

4. .... build reefs found in shallow tropical seas.

   a) Hermatypic corals b) Ahermatypes corals

   c) Both a & b d) None of these

5. .... per cent of the corals are located in the Indian ocean,Red sea and Persian Gulf.

   a) 30% b) 5%

   c) 10% d) 70%

6. Optimum temperature for coral growth is ....

   a) 23 to 25°C b) 10 to 12°C

   c) 25 to 35°C d) 5-10°C

7. Difference in coral growth between the Persian Gulf and the Gulf of Aqaba is due to…..

   a) Atoll b) Shamal

   c) Calanoids d) Lagoons

8. The growth of corals takes place in the ....

   a) Photic zone b) Aphotic zone

   c) Hypolimnion d) Metalimnion

9. Inhibits the growth of corals because corals require sediment free water.
   a) Transparency   b) Hardness
   c) Turbidity   d) All of these
10. Gives information about the history of ocean basins and sea level changes.
   a) Ocean turbidity   b) Tides
   c) Reef formation   d) Boat channel
11. Are separated from the shore by a narrow and shallow lagoon known as boat channel.
   a) Barrier reefs   b) Table reefs
   c) Fringing reefs   d) Patch reefs
12. Are the largest and the most extensive of all the reefs.
   a) Barrier reefs   b) Patch reefs
   c) Table reefs   d) Both b & c
13. Is the longest barrier reef in the world.
   a) Bahama platform
   b) The Great Barrier Reef of Australia
   c) Atolls
   d) Sunderban
14. Are small, open ocean reefs that have no central islands or lagoons.
   a) Barrier reefs   b) Patch reefs
   c) Table reefs   d) Both a & b
15. Is the largest table reef.
   a) Bahama platform   b) Carribbean
   c) Both a & b   d) None of these
16. Are small, circular or irregular reefs that rise from the floor of lagoons, behind barrier reefs or within atolls.
   a) Table reefs   b) Atolls
   c) Patch reefs   d) Barrier reefs
17. Are small annular atolls in which the lagoon is a small pool.
   a) Faros   b) Table reefs
   c) Shallow lagoon reefs   d) Coral pinnacles

18. Are chains of small atolls with lagoons of shallow depth.
    a) Patch reefs	b) Faros
    c) Coral pinnacles	d) Dana Davis
19. Darwin-Dana-Davis theory is also known as ....
    a) Subsidence theory	b) Antecedent platform theory
    c) Glacial control theory	d) Karstic saucer theory
20. The Antecedent Platform Theory was proposed by ....
    a) Darwin-Dana-Davis	b) Hoffmeister and Ladd
    c) Yabe and Asano	d) Daly
21. Karstic Saucer Theory of atoll origin was developed by ....
    a) Yabe and Asano	b) Daly
    c) Robert Clown	d) Hoffmeister and Ladd
22. The coral of each polyp is called as ....
    a) Corallum	b) Corallite
    c) Gastropores	d) Gastrozooids
23. The hydrozoan corals are known as ....
    a) Corallum	b) Hydrocorallum
    c) Atoll	d) Gastrozooids
24. Corals are produced by ....
    a) Millepora	b) Stylasterina
    c) Both a & b	d) None of these
25. Example of octocorallian coral is ....
    a) Alcyonium	b) Tubipora
    c) Helipora	d) All of these
26. Example of hexacorallian coral is ....
    a) Favea	b) Astrea
    c) Madrepora	d) All of these
27. Match the pairs:
    a) Alcyonium	i) Orange pipe coral
    b) Tubipora	ii) Sea fan
    c) Heliopora	iii) Stony coral
    d) Gorgonia	iv) Mushroom coral

e) Fungia v) Brain coral
f) Favea vi) Stag-horn coral
g) Astraea vii) Dead-man's fingers
h) Madrepora viii) Star coral
i) Meandrina ix) Blue coral

a) a=vii,b=ix,c=i,d=ii,e=iv.f=iii,g=vi,h=viii,i=v
b) a=vii,b=i,c=ix,d=ii,e=iv,f=iii,g=vi,h=iv,i=v
c) a=vii,b=i,c=ix,d=ii,e=iv,f=iii,g=viii,h=vi,i=v
d) a=vii,b=i,c=ix,d=ii,e=iv,f=viii,g=iii,h=v,i=iv

28. Which of the following is an anthozoan coral?
a) Alcyonium b) Tubipora
c) Helipora d) All of these

29. The coral is produced by ....
a) Coral polyps b) Septa
c) Theca d) Epitheca

30. Thoysands of corallites fuse together to form a large coral stone called ....
a) Cosate b) Epitheca
c) Corallum d) Columella

31. Volcanic crater theory was proposed by ....
a) Daly b) Stutchbury
c) Darwin d) Ritz

32. Benefits are coral reefs are ....
a) They protect the sea-shore from erosion
b) They provide an ideal habitat for various marine animals
c) They produce islands
d) All of these

33. In India, coral reefs are present in the areas of ....
a) Gulf of Kutch b) Gulf of Mannar
c) Andaman & Nicobar d) All of these

34. Which year was declared as the international year of reef?
a) 2010 b) 2003
c) 2008 d) 2006

35. Lakshadweep reefs are ....
    a) Atolls
    b) Fringing reefs
    c) Table reefs
    d) Patch reefs
36. Gulf of Kutch, Palk Bay and Gulf of Mannar are ....
    a) Fringing reefs
    b) Atoll
    c) Patch reefs
    d) Table reefs
37. What are the threats that push corals to the brink of extinction?
    a) Global warming
    b) Coral bleaching
    c) Marine pollution
    d) All of these
38. The baseline map of Indian coral reef is prepared by data derived from ....
    a) RESOURCESAT-1 (IRS-P6)
    b) INCOIS-2 (IRS-G6)
    c) GEOSAT-3 (IRS-2)
    d) OCEANOSAT-2
39. Coral reefs are known as ....
    a) Tropical Rainforests of the sea
    b) The medicine chests of the sea
    c) Mangroves
    d) Both a & b
40. Coral reefs have the potential to provide cures for life threatening diseases such as ....
    a) Cardio-vascular problems
    b) Ulcers
    c) Leukemia
    d) All of these
41. Bio-invasion of .... have caused detrimental impact on reef corals.
    a) Kappaphycus alvarezii
    b) Octocoral cariijoa rissei
    c) Both a & b
    d) None of these
42. Conservation of Indian coral reefs is enforced under ....
    a) Marine Fishing Regulation Act (MFRA) 2000
    b) Coastal Regulation Zone (CRZ) Notification 2011
    c) Both a & b
    d) Water Pollution Prevention Act (1976)
43. Causes extensive damage to the reef.
    a) White pox
    b) White band
    c) White plague
    d) All of these

44. What is the name of the animal part of coral reef?
    a) Polyp
    b) Zooxanthellae
    c) Pink plaque
    d) Both a & b
45. What is the name of symbiotic organism the live inside a coral?
    a) Zooxanthellae
    b) Chlorella
    c) Diatoms
    d) All of these
46. How fast do most corals grow per year?
    a) 5 inch
    b) 3 inch
    c) 2 inch
    d) 1 inch
47. For the first time indigenously developed Remotely operated vehicle (PROVE) is used for ....
    a) Mapping the depth of Indian ocean
    b) Mapping the coral reefs in Andaman & Nicobar Islands
    c) Mapping the length of coastal line
    d) All of these
48. What percent of ocean surface is covered by coral reefs?
    a) 0.5%
    b) 0.8%
    c) 0.1%
    d) 0.7%
49. Since February 2011,.................service initiated from INCOIS .
    a) Coral Bleaching Alert System
    b) Tsunami Alert System
    c) Information on Potential Fishing Zone
    d) All of these
50. Uses the satellite derived sea surface temperature (SST) in order to assess the thermal stress accumulated in the coral environs.
    a) Coral Bleaching Alert System (CBAS)
    b) Coral Density Alert System (CDAS)
    c) Both a & b
    d) None of these
51. The 'Status of Coral Reefs of the World:2020'report was produced by ....
    a) Bay of Bengal Programme-IGO
    b) Global Coral Reef Monitoring Network

c) International Coral Reef Initiative

d) United Nations Environment Programme

52. The International Coral Reef Initiative (ICRI) was founded in...........

a) 1990 b) 1994

c) 1972 d) 1999

53. The Global Coral Reef Monitoring Network (GCRMN) was established in ....

a) 1995 b) 2005

c) 1999 d) 2007

54. Intergovernmental Panel on Climate Change (IPC c)'ssixthAssessment Report on Impact,Adaptation and Vulnerability was released in ....

a) February 2023 b) February 2021

c) February 2022 d) February 2020

55. Coral reefs are home to ....

a) 4000 species of reef fish

b) 840 species of corals

c) Over 1 million species of other animals

d) All of these

56. In which year Gulf of Mannar Marine National Park was declared ?

a) 1986 b) 1996

c) 1999 d) 2006

57. Is a joint venture of Wildlife Trust of India and the Gujarat Forest Department.

a) Coral Reef Recovery Project

b) Coral Bleaching Alert System

c) Save Corals

d) Conservation of marine protected areas

58. In which year Coral Reef Recovery Project was launched ?

a) 2006 b) 2008

c) 2010 d) 2012

59. The mandates of the Coral Reef Recovery Project include ....

a) Generation of baseline data on the coral diversity in Mithapur reef

b) Generation of baseline data on the diversity of other marine life forms in Mithapur reef

c) Reintroduction of locally -extinct species

d) All of these

60. What is the proper sequence in reef formation?

a) Atoll, Barrier, Fringing reef
b) Barrier, Fringing reef, Atoll
c) Fringing reef, Barrier, Atoll
d) None of these

Answer Keys

1	a)	Animals
2	c)	Cnidaria
3	b)	Polyp
4	a)	Hermatypic corals
5	a)	30%
6	a)	23 to 25^{0}C
7	b)	Shamal
8	a)	Photic zone
9	c)	Turbidity
10	c)	Reef formation
11	c)	Fringing reefs
12	a)	Barrier reefs
13	b)	The Great Barrier Reef of Australia
14	c)	Table reefs
15	a)	Bahama platform
16	c)	Patch reefs
17	c)	Shallow lagoon reefs
18	b)	Faros
19	a)	Subsidence theory
20	b)	Hoffmeister and Ladd
21	a)	Yabe and Asano
22	b)	Corallite
23	b)	Hydrocorallum
24	c)	Both a & b
25	d)	All of these
26	d)	All of these
27	c)	a=vii,b=i,c=ix,d=ii,e=iv,f=iii,g=viii,h=vi,i=v
28	d)	All of these
29	a)	Coral polyps
30	c)	Corallum
31	b)	Stutchbury

32	d)	All of these
33	d)	All of these
34	c)	2008
35	a)	Atolls
36	a)	Fringing reefs
37	d)	All of these
38	a)	RESOURCESAT-1 (IRS-P6)
39	d)	Both a & b
40	d)	All of these
41	c)	Both a & b
42	c)	Both a & b
??	d)	Water Pollution Prevention Act (1976)
43	d)	All of these
44	a)	Polyp
45	a)	Zooxanthellae
46	d)	1 inch
47	b)	Mapping the coral reefs in Andaman & Nicobar Islands
48	c)	0.1%
49	a)	Coral Bleaching Alert System
50	a)	Coral Bleaching Alert System (CBAS)
51	b)	Global Coral Reef Monitoring Network
52	b)	1994
53	a)	1995
54	c)	February 2022
55	d)	All of these
56	a)	1986
57	a)	Coral Reef Recovery Project
58	b)	2008
59	d)	All of these
60	c)	Fringing reef, Barrier, Atoll

7

Dairy Science

1. Which period is known as the 'operation flood era'?
   a) 1971 and 1996
   b) 1978 and 1999
   c) 1969 and 1974
   d) 1974 and 1978
2. In which year Indian government launched 'operation flood program'?
   a) 1970
   b) 1950
   c) 1960
   d) 1972
3. Which is the largest milk producer in the world?
   a) China
   b) Germany
   c) Russia
   d) India
4. 'White Revolution' in concerned with ....
   a) Egg production
   b) Meat production
   c) Grain production
   d) Milk production
5. From which year National Project for Cattle and Buffalo Breeding (NPCBB) was started?
   a) 2000
   b) 1998
   c) 2011
   d) 1996
6. The mandate of NPCBB is to ....
   a) Arrange delivery of vastly improved artificial insemination service at the farmer's doorstep
   b) Bring all breed able females among cattle and buffalo under organized breeding through AI or natural service by high quality bulls
   c) Undertake breed improvement programme for indigenous cattle and baffoles so as to improve the genetic make-up as well as their availability
   d) All of these

7. .... contribute 24% of the global milk production.
   a) India b) United States
   c) Russia d) Netherlands
8. .... is widely renowned as the 'Father of White Revolution in India'.
   a) Dr.Verghese Kurian b) K.H.Alikunhi
   c) S.A.H.Abidi d) Dr.M.Sinha
9. 'World Milk Day' is observed on ....
   a) 1st June b) 2nd July
   c) 26th November d) 2nd June
10. 'World Milk Day' is observed since ....
   a) Year 2008 b) Year 2004
   c) Year 2001 d) Year 2007
11. 'National Milk Day' is observed on ....
   a) 1st June b) 26th November
   c) 4th March d) 10th November
12. On occasion of ....the government of India confer 'National Gopal Ratna Award'.
   a) National Milk Day b) World Milk Day
   c) Independence Day d) Republic day
13. The milk productivity in India is low due to ....
   a) Poor quality feed and fodder availability
   b) Improper infrastructure
   c) Poor healthcare
   d) All of these
14. Top Indian milk producing state is ....
   a) Rajasthan b) Maharashtra
   c) West Bengal d) Bihar
15. .... may be defined as milk which has been heated in sealed container continuously to a temperature of 115 °C for 15 minutes.
   a) Flavoured milk b) Pure milk
   c) Sterilized milk d) All of these

16. ....is prepared by admixture oof cow and buffalo milk.
    a) Flavoured milk
    b) Sterilized milk
    c) UHT milk
    d) Toned milk
17. ....is a cow milk or buffalo milk or sheep milk or goat milk or a combination of any one of these milk that has been standardized to fat minimum 4.5%.
    a) Recombined milk
    b) Standardized milk
    c) Skimmed milk
    d) Toned milk
18. ....is prepared from milk from which almost all the milk fat has been removed mechanically.
    a) Skimmed milk
    b) Standardized milk
    c) Recombined milk
    d) Both b & c
19. Addition of vitamins and minerals to milk is known as ....
    a) Humanized milk
    b) Fortified milk
    c) Soft -curd milk
    d) Reconstituted milk
20. .... refers to milk prepared by dispersing whole milk powder in water.
    a) Fortified milk
    b) Humanized milk
    c) Reconstituted milk
    d) Skimmed milk
21. .... is a product resembling milk but of non-dairy origin.
    a) Filled milk
    b) Imitation milk
    c) Toned milk
    d) Fortified milk
22. .... are dairy foods that have been fermented with lactic acid bacteria such as Lactobacillus and Lactococcus.
    a) Fortified milk
    b) Humanized milk
    c) Fermented milk
    d) Reconstituted milk
23. Health benefit of fermented milk is ....
    a) Much more palatable than milk
    b) Nutritive value usually increased
    c) May possess therapeutic properties
    d) All of these
24. ....is produced by adding culture named *Lactobacillus acidophilus*.
    a) Acidophilus milk
    b) Butter milk
    c) Imitation milk
    d) Filled milk

25. ....is a coagulated product obtained from pasturized or boiled milk.
    a) Fat milk
    b) Yoghurt
    c) Imitation milk
    d) All of these
26. Which of the following is dairy product?
    a) Skim milk
    b) Butter milk
    c) Ghee residue
    d) All of these
27. .... is defined as a product of commercial value produced during the manufacture of a main product.
    a) By-product
    b) Main product
    c) Commercial product
    d) Value added product
28. .... is industrial casein which has been precipitated by various acids from skim milk.
    a) Acid casein
    b) Free casein
    c) Both a & b
    d) None of these
29. Rennet casein is used for the production of ....
    a) Fibre
    b) Paint
    c) Paper coating
    d) All of these
30. .... is industrial casein which has been precipitated by rennet from skim milk.
    a) Rennet casein
    b) Greese casein
    c) Fibril casein
    d) Both b & c
31. .... is used in various food products such as ice cream and coffee.
    a) Rennet casein
    b) Edible casein
    c) Both a & b
    d) None of these
32. .... is a nourishing soft drink.
    a) Whevit
    b) Velvet
    c) Welex
    d) Ventex
33. 'Whevit' is developed by ....
    a) National Dairy Research Institute
    b) National Bureau of Animal Genetic Resources
    c) Central Institute for Research on Buffaloes
    d) National Milk Grid

34. .... is obtained as a by-product of channa/paneer/cheese industries.
    a) Whey b) Ghuva
    c) R-casein d) Both a & b

35. .... is a food safety program that was developed for NASA.
    a) CAC b) ISO
    c) HACCP d) GMP

36. Good manufacturing Practice (GMP) deals with ....
    a) Contamination by people
    b) Contamination by food materials
    c) Contamination by packaging materials
    d) All of these

37. ....ensures the safety and quality of products to protect consumers worldwide.
    a) WHO standards b) ISO standards
    c) FAO standards d) FSSAI standards

38. .... has been established under Food Safety and Standards Act,2006.
    a) FSSAI standards b) ISO standards
    c) CAC standards d) HACCP

39. In how many phases operation flood was implemented?
    a) Two b) Four
    c) Three d) Five

40. AGMARK is a ....
    a) Quality certification mark
    b) It ensures quality and purity of a product
    c) It acts as a third-party guarantee to quality certified
    d) All of these

41. Functions performed by FSSAI are ....
    a) Framing of regulations to lay down the standards and guidelines in relation to articles of food
    b) Promote general awareness about food safety and food standards
    c) Provide training programmes for persons who are involved or intend to get involved in food business
    d) All of these

42. Milk packaging plastic materials are made up of……
    a) Low Density Polyethylene (LDPE)
    b) Polyolefins
    c) High Density Polyethylene
    d) All of these
43. The function of a dairy cooperative society is…………..
    a) Collection of milk from farmer
    b) Make regular payment to suppliers
    c) Dispatch the milk collected to milk union
    d) All of these
44. In which year Intensive Cattle Development Project (ICDP) was launched?
    a) Year 1972-73 b) Year 1964-65
    c) Year 1967-68 d) Year 1959-60
45. Intensive Cattle Development Project (ICDP) was launched with the objectives of……
    a) To provide credit facilities for the purchase of milch animals
    b) To provide dairy extension services to make farmers more interested in scientific practices of livestock farming and to increase production
    c) To provide effective marketing facilities for the milk produced by farmers through establishment of cooperative societies
    d) All of these
46. Objectives of 'Operation Flood' were……..
    a) Increase milk production
    b) Augmenting rural incomes
    c) Reasonable prices for consumers and producers
    d) All of these
47. In which year the National Dairy Development Board was established?
    a) 1971 b) 1965
    c) 1968 d) 1963
48. The headquarters of NDDB is situated in ....
    a) Mumbai (Maharashtra) b) Karnal (Haryana)
    c) Anand (Gujarat) d) Bangalore (Karnataka)

49. Objectives of NDDB are ....
    a) To direst India's dairy industry development
    b) To plan and provide farmer extension services
    c) To improve dairy technologies
    d) All of these

50. .... was the founder chairman of the NDDB
    a) Dr.Verghese Kurian
    b) Tribhuvandas Patel
    c) Dr.K.H.Alikunhi
    d) Prof.V.Sarvapalli

51. The Indian Veterinary Research Institute is situated in ....
    a) Pune
    b) Hyderabad
    c) Tatanagar
    d) Izatnagar

52. In which year the Indian Veterinary Research Institute was established?
    a) 1887
    b) 1979
    c) 1889
    d) 1979

53. The old name of National Dairy Research Institute was ....
    a) Imperial Dairy Institute
    b) Imperial Institute of Animal Husbandry and Dairying
    c) Indian Dairy Institute
    d) Both a & b

54. NARARD works for ....
    a) Providing refinance to lending institutions in rural areas
    b) Bringing about or promoting institutional development
    c) Evaluating, monitoring and inspecting the client banks
    d) All of these

55. International Dairy Federation was established in ....
    a) 1910
    b) 1903
    c) 1918
    d) 1928

56. .... was established under operation flood programme.
    a) NABARD
    b) National Milk Grid
    c) IVRI
    d) IDI

57. Headquarters of National Bureau of Animal Genetic Resources is at ....

a) Delhi
b) Hissar
c) Karnal
d) Pusa

58. 'Mission Towards Zero Non-descript AnGR of India' was launched in..........

a) Year 2019
b) Year 2010
c) Year 2009
d) Year 2021

59. 'Rashtriya Gokul Mission Scheme' was launched in ....

a) 2014
b) 2010
c) 2020
d) 2019

60. .... was launched with the objective of promoting the conservation and development of indigenous breeds of cattle and buffaloes.

a) Rashtriya Goshala Mission
b) Rashtriya Gokul Mission
c) National Dairy Project in North-East India
d) Rajiv Gandhi Scheme for Central India

61. The National Programme for Bovine Breeding and Dairy Development was launched in ....

a) 2011
b) 2017
c) 2014
d) 2002

62. Which of the following is originated from Montgomery (Pakistan) and mainly found in Bihar, Haryana and Uttar Pradesh.

a) Sahiwal
b) Gir
c) Jersy
d) Karan swiss

63. Home tract of Red Sindhi is ....

a) Kathiawar (Gujarat)
b) Karanchi (Pakistan)
c) Hisar (Haryana)
d) Devni (Maharashtra)

64. Which of the following is known as 'Black gold' or 'Pride of Haryana'?

a) Murrah
b) Nili Ran
c) Surti
d) Devni

65. Home tract of.........is Kaira and Baroda districts of Gujarat.

a) Jaffarabadi
b) Bhadawari
c) Surati
d) Mehsana

66. Home tract of Jaffarabadi and Mehsana is ....
    a) Maharashtra b) Gujarat
    c) Tamil Nadu d) Goa
67. ....are known as 'poor man's cow'?
    a) Goats b) Buffaloes
    c) Camels d) None of these
68 . ... is found in Agra, Mathura and Etawah districts of Uttar Pradesh.
    a) Beetle b) Setal
    c) Jamunapari d) Both a & b
69. Beetle mainly found in ....
    a) Andhra Pradesh and Maharashtra b) West Bengal and Orissa
    c) Bihar and Orissa d) Punjab and Haryana
70. ....is found near Behror in Alwar district of Rajasthan.
    a) Jakhrana b) Beetle
    c) Osmanabadi d) Barbari
71. ....is known for its muscular build, strength and is native to Andhra Pradesh region.
    a) Sahiwal b) Ongole
    c) Gir d) Red Sindhi
72. ....is originated in the Netherlands and is often used in commercial dairy farming in India.
    a) Ongole b) Holstein Friesian
    c) Ayrshire d) Guernsy
73. ....originated in Scotland and is often used for commercial dairy farming in India.
    a) Ayrshire b) Dozi
    c) Ongole d) Jersy
74. The Indian breed of cattle with long pendulous ears resembling a curled leaf is ....
    a) Sahiwal b) Red sindhi
    c) Gir d) Deoni
75. Which of the following exotic cattle breed was not used in the cross breeding programme in Kerala?
    a) Ayrshire b) Jersey
    c) Brown swirls d) Holstein Friesian

76. The important breeds of goats used for milk include ....

a) Barbari b) Malabari

c) Sirohi d) All of these

77. Which of the following breed is from Maharashtra?

a) Jakhrana b) Surti

c) Osmanabadi d) Malabari

78. ....is the technique through which semen are deposited into female reproductive tract by mechanical means with help of an instrument.

a) Breeding b) Artificial insemination

c) Boar d) Bundh breeding

79. Match the pairs:

a) Bull i) Adult male goat

b) Buck ii) Female parent

c) Boar iii) Adult female goat

d) Dam iv) Adult male cattle

e) Doe v) Adult female horse

f) Mare vi) Adult male pig

a) a=iv,b=i,c=vi,d=ii,e=iii,f=v b) a=iv,b=ii,c=iii,d=v,e=i,f=v

c) a=iv,b=i,c=vi,d=iii,e=ii,f=v d) a=iv,b=i,c=vi,d=ii,e=v,f=iii

80. Term used for species of buffalo and cattle is ....

a) Doe b) Buck

c) Bovine d) Culling

81. The process of elimination of non-productive or undesirable animals is known as ....

a) Castration b) Culling

c) Caprine d) Colostrum

82. ....is highly nutritious and a rich source of antibodies.

a) Caprine b) Boar

c) Colostrum d) Browse

83. Scientific name of 'domestic' or 'water buffalo' is ....

a) Sus domesticus b) Sus vittatus

c) Mystus vittatus d) Bubalus bubalis

84. Which of the following is a breed of Indian buffaloes?
   a) Nagpuri b) Toda
   c) Kundi d) All of these
85. .... has been developed from crosses between murrah and surti.
   a) Tarai b) Toda
   c) Mehsana d) Kundi
86. .... is a breed evolved through grading up of local buffaloes of coastal Andhra Pradesh with Murrah over generations.
   a) Godavari b) Nili-Ravi
   c) Manda d) Toda
87. The Nili and Ravi are types of ....
   a) Goat b) Buffaloes
   c) Camels d) None of these
88. .... are seen in their purest form in the Gir forest of Kathiawar.
   a) Bhadawari b) Manda
   c) Jaffarabadi d) Kalhandi
89. .... is found in Maharashtra.
   a) Nagpuri b) Panndhepuri
   c) Both a & b d) Kalhandi
90. .... is a buffalo breed distributed in Orissa and Andhra Pradesh.
   a) Toda b) Jerangi
   c) Nagpuri d) Gobies
91. Dairy industry in India is mainly ....
   a) Goat oriented b) Buffalo oriented
   c) Both a & b d) None of these
92. .... contributes more than 55% of total milk production of India.
   a) Buffalo milk b) Goat milk
   c) Both a & b d) None of these
93. More than 80% of buffalo cows in India calve during the period of ....
   a) March to April b) July to December
   c) January to May d) November to March
94. Buffalo calving are minimum in ....
   a) Cold months b) Rainy months
   c) Hotter months d) Both a & b

95. Which factor influence the annual calving pattern?
    a) Rainfall b) Ambient temperature
    c) Photoperiod d) All of these
96. Buffalo is………
    a) Monestrous b) Polyestrous
    c) Postpartum estrous d) Both b & c
97. Animals that cycle continuously throughout the year is known as……..
    a) Postpartum estrous animals b) Monoestrous animals
    c) Polyestrous animals d) None of these
98. Estrous cycle generally starts ....
    a) After the birth
    b) After sexual maturity of male
    c) After sexual maturity of female
    d) Both b &c
99. In buffalo the estrous cycle is about ....
    a) 52 days b) 21 days
    c) 8 days d) 38 days
100. Which changes occur during pro-estrous ?
    a) FSH hormone stimulates the growth of graafian follicles
    b) Vula starts swelling
    c) Mucus secretion
    d) All of these
101. In buffalo in pro-estrous may continue for……
    a) 10-12 days b) 2-3 days
    c) 15-17 days d) 7-8 days
102. Which period is know as 'heat period'?
    a) Estrous b) Meta-estrous
    c) Pro-estrous d) Pro-estrous
103. Which changes occur in last part of estrous?
    a) Secretion of LH increases b) Secretion of FSH decreases
    c) Ovulation occurs d) All of these

104. The last phase of estrous cycle is ....

a) Meta-estrous b) Dio-estrous

c) Polyestrous d) Monoestrous

105. Dio-estrous lasts for ....

a) 5-8 days b) 2-3 days

c) 10-18 days d) 27-30 days

106. Bang's disease is caused by ....

a) Mycobaterium bovis b) Theleria parva

c) Brucella abortus d) None of these

107. Preganancy diagnosis in buffaloes is done by ....

a) Rectal palpation b) Hormone assays

c) Both a & b d) None of these

108. In buffaloes.........is a simple, economic and most widely practiced method for pregnancy diagnosis?

a) Palpation per rectum b) FSH concentration in milk

c) LH concentration in milk d) Both b & c

109. .... method is only accurate from day 45 of pregnancy.

a) Blood plasma test b) Urine test

c) Palpation per rectum d) Both a &b

110. In buffalo, the gestation period varies from ....

a) 6 weeks less than a year b) 3 weeks to 5 weeks

c) 2-3 weeks d) 13-16 weeks

111. Buffalo pregnancy can be diagnosed with help of ....

a) Progesterone concentration in milk b) Blood plasma

c) Both a & b d) None of these

112. What is parturition?

a) The action of giving birth to young

b) The action of egg laying

c) Attaining sexual maturity

d) Both a & b

113. In the peak breeding period of buffaloes ....

a) FSH concentration is highest b) LH concentration is lowest

c) Both a & b d) FSH & LH concentration remains constant

114. adversely affect libido and semen production.

a) High ambient temperature b) Direct solar radiation

c) Both a & b d) None of these

115. The major problem of male fertility is ....

a) Poor libido

b) Quantitative reduction in buffalo semen during summer

c) Qualitative reduction in buffalo semen during summer

d) All of these

116. What is artificial insemination?

a) Semen is collected artificially and deposited in the uterus of cow

b) Semen is collected and kept in straws

c) Semen is collected and stored

d) Semen is washed and properly stored

117. BAIF Development Research Foundation is located in ....

a) Bangalore b) Mysore

c) Pune d) Bhopal

118. In which year BAIF Development Research Foundation was established?

a) 1962 b) 1967

c) 1982 d) 1992

119. Reasons for better utilization of nutrients in buffaloes are ....

a) Large rumen volume b) High rate of salvination

c) Slow rumen motility d) All of these

120. The microbial population in rumen of buffalo includes ....

a) Protozoa b) Total viable bacteria

c) Amylolytic bacteria d) All of these

121. Major cause of calf mortality is ....

a) Ascariasis b) Neonatal diarrhoea

c) Pneumonia d) All of these

122. Central Institute for Research on Buffaloes is located at ....

a) Pune (Maharashtra) b) Bhopal (Madhya Pradesh)

c) Hisar (Haryana) d) Kochi (Kerala)

123. Headquarters of National Bureau of Animal Genetic Resources is at ....

a) New Delhi
b) Karnal
c) Noida
d) Pune

124. Milk production in India has been increasing with an annual growth rate of ....

a) 2%
b) 8%
c) 6%
d) 1.5%

125. Buffaloes account for around.........of the total milk production in India.

a) 56%
b) 44%
c) 22%
d) 89%

126. Cows account for around ....of the total milk production in India.

a) 22%
b) 56%
c) 44%
d) 28%

127. ....procure milk from small -scale farmers and process it into various milk products.

a) Amul Dairy
b) Mother Dairy
c) Both a & b
d) None of these

128. Which barns are needed for proper housing of different classes of dairy stock?

a) Cow houses
b) Isolation box
c) Calving box
d) All of these

129. Advantages of tail to tail system are ....

a) Lesser danger of spread of diseases from animal to animal
b) Cow can always get more fresh air from outside
c) In cleaning middle alley is of great advantage
d) All of these

130. Advantage of face to face system are ....

a) Cows make a better showing for visitors when heads are together
b) The cows feel easier to get into their stalls
c) Sun rays shine in the gutter where they are needed most
d) All of these

131. ....is a type of dairy barn stall.
   a) The stanchion stall b) The tie stall
   c) Both a & b d) None of these

132. .... requires a few inches longer and wider than the stanchion stall.
   a) The Tie stall b) The Bowl stall
   c) The Nessel stall d) Both b & c

133. Animals suffering from infectious diseases must be kept in ....
   a) Calving boxes b) Isolation boxes
   c) Bull boxes d) Bullock shad

134. Cattles can be marked by ....
   a) Ear-tag b) Tatto
   c) Number tags d) All of these

135. Advantages of weaning are ....
   a) Milking without a calf is more hygienic and sanitary
   b) The calf can be culled out at any early age
   c) Total yield increases
   d) All of these

136. If the owner wants milk in quantity, he must practice ....
   a) Brining system b) Weaning system
   c) Branding d) Notching

137. Advantages of dehorning are ....
   a) Dehorned animal will need less space in the sheds
   b) Horned animals are a danger to the operator
   c) Dehorned animals can be handled more easily
   d) All of these

138. Mechanical method of dehorning includes.....
   a) Use of electrical dehorners b) Use of rubber bands
   c) Use of clippers and saws d) All of these

139. .... is the unsexing of the male or female and consists in the removal of both testicles or ovaries respectively.
   a) Dehorning b) Castrating
   c) Horning d) None of these

140. Castration is done with the help of ....

a) Castration clip b) Burdizzo's castrator

c) Klinner's instrument d) Glimm's clip

141. Castration with the help of Burdizzo's castrator is known as ....

a) Bloodless castration b) Rough castration

c) False castration d) Both a & b

142. In which state Indian goat, Gaddi is distributed?

a) Himachal Pradesh b) Uttar Pradesh

c) Orissa d) Both a & b

143. ....is also known as Pashmina.

a) Gallti b) Sulari

c) Gurri d) Changthangi

144. .... are mostly reared in Ladakh and Spiti valleys and its neighbouring areas of Himachal Pradesh.

a) Kutchi b) Changthangi

c) Sirohi d) Both b & c

145. Advantages of sheep farming are ....

a) Sheep hardly damage any tree

b) Sheep are of economical convertor of grass into meat and wool.

c) Sheep do not need expensive buildings to house them

d) All of these

146. Which of the following sheep breed is native of Spain?

a) Lincoln b) Corriedale

c) Leicester d) Merinos

147. Lincoln breed is native of ....

a) England b) Spain

c) Pakistan d) India

148. 'Indian Dairyman' is published by ....

a) Indian Dairy Association

b) Indian Society of Agricultural Economics

c) Indian Council of Agricultural Research

d) Association of Food Scientists and Technologists

149. 'Indian Journal of Dairy Science' is published by ....
   a) Indian Society of Animal Science
   b) Animal Nutrition Association
   c) Indian Dairy Association
   d) Association of Food Scientists and Technologists
150. Which of the following bill was introduced in Loksabha in August 2005?
   a) Dairy Products Standards Bill 2005
   b) Dairy Standards Bill 2005
   c) Food Safety and Standards Bill 2005
   d) Dairy Based Products Bill 2005
151. Which disease is caused by the acid-fast bacteria Mycobacterium bovis?
   a) Mastitis b) Bovine tuberculosis
   c) Ich disease d) Both a & b
152. .... is multi- factorial disease predominantly due to management related problems.
   a) Nocardiasis b) Mastitis
   c) Coccidiosis d) Dropsy
153. .... is a haemoprotozoan disease.
   a) Theileriosis b) Dropsy
   c) Ich disease d) Both a & b
154...... is used for treatment of theileria infected animals.
   a) Penicillin b) Acetyl chloride
   c) Ascorbic acid d) Tetracyclines
155. .... is caused by an apthovirus belonging to the family picornaviridae.
   a) Foot and mouth disease b) Leptospirosis
   c) Coccidiosis d) None of these
156. Which disease is related with reproductive problems and milk drop syndrome?
   a) Leptospirosis b) Brucellosis
   c) Bovine tuberculosis d) Both b & c
157. Match the pairs:
   a) Khillari i) Karnataka
   b) Kherigarh ii) Tamil Nadu

c) Bachaur — iii) Madhya Pradesh
d) Kenkatha — iv) Uttar Pradesh
e) Amritmahal — v) Maharashtra
f) Nilliravi — vi) Punjab

a) a-ii,b-iv,c-i,d-iii,e-v,f-vi
b) a-v,b-iv,c-ii,d-iii,e-i,f-vi
c) a-v,b-iv,c-i,d-iii,e-ii,f-vi
d) a-v,b-iv,c-i,d-iii,e-vi,f-ii

158. Anthrax is also known as ....
a) Splenic fever
b) Pasteurellosis
c) Shipping fever
d) Black quarter

159. Which of the following disease is bacterial disease?
a) Brucellosis
b) Pneumonia
c) Babesiosis
d) Both a & b

160. Rinerpest is a ....
a) Viral infection
b) Protozoan disease
c) Worm disease
d) Bacterial disease

161. Which of the following is protozoan disease?
a) Babesiosis
b) Theileriosis
c) Trichomonads
d) All of these

162. Which of the following are non-specific disorders?
a) Bloat
b) Ketosis
c) Milk fever
d) All of these

163. Which is causative agent of Splenic fever?
a) Tape worm
b) Liver fluke
c) Bacillus anthracis
d) Pasteurella multocida

164. Anthrax is characterized by ....
a) Foamy blood from nose and mouth
b) Sudden rise in temperature
c) Abortion in cows
d) All of these

165. Penicillin is used for treatment of ....
a) Anthrax
b) Pasteurellosis
c) Tuberculosis
d) All of these

166. Haemorrhagic septicemia is caused by ....
   a) Costia necatrix
   b) Pasteurella multocida
   c) Eimeria necatrix
   d) None of these
167. Bain's vaccine is used for control of ....
   a) Black-quarter
   b) Haemorrhagic septicemia
   c) Theileriosis
   d) Trypanosomiasis
168. Black-quarter is caused by ....
   a) Clostridium chauvoei
   b) Brucella abortus
   c) Eimeria necatrix
   d) Eimeria tenella
169. Contagious bovine abortions also known as ....
   a) Bang's disease
   b) Brucellosis
   c) Tuberculosis
   d) Both a & b
169. Bang's disease is caused by ....
   a) Mycobaterium bovis
   b) Theleria parva
   c) Brucella abortus
   d) None of these
170. Which drugs are used for treatment of Brucellosis?
   a) Streptomycin
   b) Aureomycin
   c) Both a & b
   d) Formalin
171. Johne's disease is known as ....
   a) Para tuberculosis
   b) Mastitis
   c) Calf sacours
   d) Cattle plague
172. Johne's disease is caused by ....
   a) Myxobolus cerebralis
   b) Mycobacterium paratuberculosis
   c) Nosema spp
   d) None of these
173. Mastitis is caused by ....
   a) Streptococcus agalactiae
   b) Staphylococcus areus
   c) Both a & b
   d) Eimeria necatrix
174. Acute mastitis is characterized by ....
   a) Repeated mild attacks of mammary swelling
   b) Production of clotted milk
   c) High fever and anorexia
   d) All of these

175. Rinder pest is commonly known as ....

a) Apthous fever
b) Cattle plague
c) Picoma
d) Mastitis

176. Mode of transmission for Rinder pest is ....

a) By direct contact of susceptible animals with infected ones by method of inhalation
b) By ingestion of infected food
c) Both a & b
d) None of these

177. Diagnosis for Bovine typhus includes ....

a) Isolation and identification of virus
b) Serum neutralization test
c) ELISA
d) All of these

178. Lumpy Skin Disease is an infectious .... of cattle.

a) Bacterial disease
b) Protozoan disease
c) Viral disease
d) Fungal disease

179. Lumpy skin disease is characterized by ....

a) High fever
b) Enlarged superficial nodes on the skin
c) Peculiar multiple lumps on the skin
d) All of these

180. LSD virus is also known as ....

a) Neethling virus
b) Ades virus
c) Rhacocephalus virus
d) Xenivirus

181. In which year Lumpy skin disease was first reported in India?

a) 2017
b) 2020
c) 2019
d) 2021

182. Lumpy skin disease is transmitted by ....

a) Flies
b) Mosquitoes
c) Ticks
d) All of these

183. The act of stopping the lactation of a cow in preparation for her next lactation is ....

a) Flushing
b) Grading up
c) Steaming up
d) Drying off

184. The most ideal method of milking is ....

a) Stripping
b) Knuckling
c) Full hand milking
d) Thumping

185. Which method is not recommended in hand milking?

a) Dry hand milking
b) Stripping
c) Full hand milking
d) Knuckling

186. The period between one calving and next conception in a cow is known as ....

a) Intercalving period
b) Service period
c) Puberty
d) None of these

187. Hormone responsible for letting down of milk is ....

a) Relaxin
b) Prolactin
c) Oxytocin
d) Inhibin

188. The gestation period of dairy cattle is ....

a) 150 days
b) 280 days
c) 139 days
d) 90 days

189. The test used for checking presence of buffalo milk in cow milk is ....

a) Baudoin test
b) California mastitis test
c) Hansa test
d) Hotis test

190. The late blowing in milk product is due to ....

a) E.coli
b) Salmonelle typhi
c) Clostridium sp
d) Aspergillus sp

191. Sterlized milk can be stored at least for ....

a) One day
b) One week
c) One month
d) One year

192. Percentage of mineral matter in milk is ....

a) 2.5%
b) 0.7%
c) 5.2%
d) 0.2%

193. Which governing body sets the limit of fat percent in different types of milk available in the market?

a) ISI b) WHO

c) FSSAI d) FCI

194. The audio autobiography of Dr. Varghese Kurien is ....

a) The man who made the elephant dance

b) I too had a dream

c) My Dairy Diary

d) The man with the billion little idea

195. Which among the following is lower in buffalo milk as compared to cow milk?

a) pH b) Viscosity

c) Curd tension d) Rennet clotting time

196. Which is not a rapid platform test performed for raw milk?

a) Direct microscopic count b) Clot on boiling point

c) Standard plate count d) Titratable acidity

197. The freezing point of cow milk is....

a) 0.555°C b) 0.560°C

c) 0.575°C d) 0.588°C

198. Objective of pasteurization is ....

a) To reduce the shelf life

b) To increase enzyme activity

c) To kill most of the bacteria which are disease causing

d) To increase the fat content of the milk

199. By which instrument water adulteration in milk is detected?

a) Refractometer b) Lactometer

c) Viscometer d) Hygrometer

200. pH of milk is a ....

a) 6.9 to 10.5 b) 10.5 to 11.7

c) 6.4 to 6.9 d) 3.4 to 5.3

Answer Keys

1	a)	1971 and 1996
2	c)	1960
3	d)	India
4	d)	Milk production
5	a)	2000
6	d)	All of these
7	a)	India
8	a)	Dr.Verghese Kurian
9	a)	1st June
10	c)	Year 2001
11	b)	26th November
12	a)	National Milk Day
13	d)	All of these
14	a)	Rajasthan
15	c)	Sterilized milk
16	d)	Toned milk
17	b)	Standardized milk
18	a)	Skimmed milk
19	b)	Fortified milk
20	c)	Reconstituted milk
21	b)	Imitation milk
22	c)	Fermented milk
23	d)	All of these
24	a)	Acidophilus milk
25	b)	Yoghurt
26	d)	All of these
27	a)	By-product
28	c)	Both a & b
29	d)	All of these
30	a)	Rennet casein
31	b)	Edible casein
32	a)	Whevit
33	a)	National Dairy Research Institute
34	a)	Whey
35	c)	HACCP
36	d)	All of these
37	b)	ISO standards
38	a)	FSSAI standards

39	c)	Three
40	d)	All of these
41	d)	All of these
42	d)	All of these
43	d)	All of these
44	b)	Year 1964-65
45	d)	All of these
46	d)	All of these
47	b)	1965
48	c)	Anand (Gujarat)
49	d)	All of these
50	a)	Dr.Verghese Kurian
51	d)	Izatnagar
52	c)	1889
53	d)	Both a & b
54	d)	All of these
55	b)	1903
56	b)	National Milk Grid
57	c)	Karnal
58	d)	Year 2021
59	a)	2014
60	b)	Rashtriya Gokul Mission
61	c)	2014
62	a)	Sahiwal
63	b)	Karanchi (Pakistan)
64	a)	Murrah
65	c)	Surati
66	b)	Gujarat
67	a)	Goats
68	c)	Jamunapari
69	d)	Punjab and Haryana
70	a)	Jakhrana
71	b)	Ongole
72	b)	Holstein Friesian
73	a)	Ayrshire
74	c)	Gir
75	a)	Ayrshire
76	d)	All of these
77	c)	Osmanabadi

78	b)	Artificial insemination
79	a)	a=iv,b=i,c=vi,d=ii,e=iii,f=v
80	c)	Bovine
81	b)	Culling
82	c)	Colostrum
83	d)	Bubalus bubalis
84	d)	All of these
85	c)	Mehsana
86	a)	Godavari
87	b)	Buffaloes
88	c)	Jaffarabadi
89	c)	Both a & b
90	b)	Jerangi
91	b)	Buffalo oriented
92	a)	Buffalo milk
93	b)	July to December
94	c)	Hotter months
95	d)	All of these
96	b)	Polyestrous
97	c)	Polyestrous animals
98	c)	After sexual maturity of female
99	b)	21 days
100	d)	All of these
101	b)	2-3 days
102	a)	Estrous
103	d)	All of these
104	b)	Dio-estrous
105	c)	10-18 days
106	c)	Brucella abortus
107	c)	Both a & b
108	a)	Palpation per rectum
109	c)	Palpation per rectum
110	a)	6 weeks less than a year
111	c)	Both a & b
112	a)	The action of giving birth to young
113	c)	Both a & b
114	c)	Both a & b
115	d)	All of these
116	a)	Semen is collected artificially and deposited in the uterus of cow

117	c)	Pune
118	b)	1967
119	d)	All of these
120	d)	All of these
121	d)	All of these
122	c)	Hisar (Haryana)
123	b)	Karnal
124	c)	6%
125	a)	56%
126	c)	44%
127	c)	Both a & b
128	d)	All of these
129	d)	All of these
130	d)	All of these
131	c)	Both a & b
132	a)	The Tie stall
133	b)	Isolation boxes
134	d)	All of these
135	d)	All of these
136	b)	Weaning system
137	d)	All of these
138	d)	All of these
139	b)	Castrating
140	b)	Burdizzo's castrator
141	a)	Bloodless castration
142	d)	Both a & b
143	d)	Changthangi
144	b)	Changthangi
145	d)	All of these
146	d)	Merinos
147	a)	England
148	a)	Indian Dairy Association
149	c)	Indian Dairy Association
150	c)	Food Safety and Standards Bill 2005
151	b)	Bovine tuberculosis
152	b)	Mastitis
153	a)	Theileriosis
154	d)	Tetracyclines
155	a)	Foot and mouth disease

156	a)	Leptospirosis
157	b)	a-v,b-iv,c-ii,d-iii,e-i,f-vi
158	a)	Splenic fever
159	d)	Both a & b
160	a)	Viral infection
161	d)	All of these
162	d)	All of these
163	c)	Bacillus anthracis
164	d)	All of these
165	a)	Anthrax
166	b)	Pasteurella multocida
167	b)	Haemorrhagic septicemia
168	a)	Clostridium chauvoei
169	d)	Both a & b
170	c)	Both a & b
171	a)	Para tuberculosis
172	b)	Mycobacterium paratuberculosis
173	c)	Both a & b
174	d)	All of these
175	b)	Cattle plague
176	c)	Both a & b
177	d)	All of these
178	c)	Viral disease
179	d)	All of these
180	a)	Neethling virus
181	c)	2019
182	d)	All of these
183	d)	Drying off
184	c)	Full hand milking
185	d)	Knuckling
186	b)	Service period
187	c)	Oxytocin
188	b)	280 days
189	b)	California mastitis test
190	c)	Clostridium sp
191	b)	One week
192	b)	0.7%
193	c)	FSSAI
194	a)	The man who made the elephant dance

195	d)	Rennet clotting time
196	c)	Standard plate count
197	a)	0.555°C
198	c)	To kill most of the bacteria which are disease causing
199	b)	Lactometer
200	c)	6.4 to 6.9

Table 1: Expansions of acronyms

Sr.No.	Acronym	Expansion
1	APEDA	Agricultural and Processed Food Products Export Development Authority
2	ANA	Animal Nutrition Association
3	BVD	Bovine Viral Diarrhoea
4	CAGR	Compound Annual Growth Rate
5	CFSP&TI	Central Frozen Semem Production & Training Institute
6	CFTRI	Central Food Technological Research Institute
7	CIRB	Central Institute for Research on Buffaloes
8	DEDS	Dairy Entrepreneurship Development Scheme
9	DFRL	Defence Food Research Laboratory
10	DHIA	Dairy Herd Improvement Association
11	FICSI	Food Industry Capacity and Skill Initiative
12	FMD	Foot and Mouth Disease
13	FSSAI	Food Safety and Standards Authority of India
14	GCMMF	Gujarat Cooperative Milk Marketing Federation Ltd.
15	GMP	Good Manufacturing Practice
16	IBR	Infectious Bovine Rhinotracheitis
17	ICDP	Intensive Cattle Development Project
18	IDEA	India Dairy Engineers Association
19	IGA	International Goat Association
20	ILRI	International Livestock Research Institute
21	IMARC	International Market Analysis Research and Consulting Group
22	ISO	International Organization for Standardization
23	LSD	Lumpy Skin Disease
24	KCMMF	Kerala Cooperative Milk Marketing Federation
25	KMF	Karnataka Cooperative Milk Producers Federation Ltd.
26	MMP	Milk and Milk Products
27	NABARD	National Bank for Agriculture and Rural Development
28	NAFED	National Agricultural Cooperative Marketing Federation of India Ltd.
29	NBAGR	National Bureau of Animal Genetic Resources

30	NCDFI	National Cooperative Dairy Federation of India Ltd.
31	NDDB	The National Dairy Development Board
32	NDP	National Dairy Plan
33	NDRI	National Dairy Research Institute
34	NMG	National Milk Grid
35	NNMB	National Nutrition Monitoring Bureau
36	NPCBB	National Project for Cattle and Buffalo Breeding
37	NPDD	National Programme for Dairy Development
38	PFA	Prevention of Food Adulteration
39	RAFAL	Regional Feed Analytical Laboratories
40	VAP	Value Added Products

8

Poultry Keeping

1. Rearing of fowls for eggs and flesh is known as ....
   a) Poultry keeping b) Husbandry
   c) Sericulture d) Apiculture
2. The types of poultry birds are ....
   a) Broilers b) Layers
   c) General purpose birds d) All of these
3. The layers are reared mainly for ....
   a) Flesh b) Eggs
   c) For both d) None of these
4. Which of the following are layers?
   a) Anconas b) Australorp
   c) White leghorn d) All of these
5. Broilers are reared for ....
   a) Flesh b) Fish
   c) Ornamental purpose d) All of these
6. Broilers are also known as ....
   a) Table breeds b) Spoon breeds
   c) Cockos d) Anconas
7. Example of broiler is ....
   a) Dorking b) Sussex
   c) Both a & b d) Anconas
8. .... Are reared for both flesh and eggs.
   a) Layers b) Broilers
   c) General purpose birds d) All of these
9. Example of general purpose bird is ....
   a) Australorp b) Sussex
   c) Rhode Island Reds d) All of these

10. The egg contains ..... .
    a) 10% shell
    b) 60% white
    c) 30% yellow yolk
    d)All of these
11. White of the egg is a ..... .
    a) Fat
    b) Protein
    c) Vitamin
    d) Carbohydrate
12. Egg contains ..... .
    a) 7.9gms protein
    b) 7.9gms fat
    c) 36gmg calcium
    d) All of these
13. The egg contains ..... .
    a) 1.62mg iron
    b) 32mg Phosphorus
    c) 103Kcal energy
    d) All of these
14. Eggs contain which vitamin?
    a) A & D
    b) E & K
    c) B
    d) All of these
15. Which fowls are good layers?
    a) Ancona
    b) Black Minorca
    c) White leghorn
    d) All of these
16. Layers show which characters?
    a) They produce maximum number of eggs
    b) They are non-broody
    c) They do not waster time by sitting on the eggs
    d) All of these
17. White leghorn originated in ..... .
    a) Australia
    b) Italy
    c) Germany
    d) India
18. Black minora is ..... .
    a) Reared for eggs
    b) Black coloured eggs
    c) Brown coloured eggs
    d) Both a & b
19. .... is reared for flesh as well as eggs.
    a) Black monora
    b) White leghorn
    c) Australorp
    d) All of these

20. The one day old chicks are sexed by ....
    a) Feather sexing b) Colour sexing
    c) Vent sexing d) All of these
21. Identification of female chicks by seeing the ccoaca is known as ....
    a) Ccoacal sexing b) Vent sexing
    c) Stripping d) Feather sexing
22. Muscular papilla is absent in the ccoaca of ....
    a) Male b) Female
    c) Both d) None of these
23. Vent sexing is a .... method.
    a) Australian b) Chinese
    c) Indian d) Japanese
24. Vent sexing was proposed by ..... .
    a) Dr. Kizoshi Masui (1927) b) Bobert Seebling (1931)
    c) Carles Deno (1942) d) Dr. Krits Fritz (1927)
25. When the cock is red coloured all the female chicks will be ..... .
    a) Red b) White
    c) Brown d) Yellow
26. The female chicks have ..... .
    a) Long primary feathers b) Short coverts
    c) Both a & b d) Long coverts
27. The male chicks have ....
    a) Short coverts b) Long coverts
    c) Short primary feathers d) Both b & c
28. Advantage of sexing is ....
    a) It avoids unnecessary feeding of male chicks
    b) It provides only female chicks for the farmers who want to rear layers
    c) Both a & b
    d) None of these
29. The type of poultry house is ....
    a) Cage house b) Layer house
    c) Deep litter house d) All of these

30. The shed in which the fowls are reared is known as ..... .
    a) Poultry house
    b) Grower house
    c) Pen house
    d) Basa house
31. ..... contains litter materials on the floor.
    a) Broiler house
    b) Brooder house
    c) Grower house
    d) Deep litter house
32. The litter materials used for deep litter house is ..... .
    a) Saw dust
    b) Wood shavings
    c) Paddy house
    d) All of these
33. The layer house is meant for ..... .
    a) Rearing of layers
    b) Rearing of broilers
    c) Both a & b
    d) None of these
34. The broiler house is meant for ..... .
    a) Rearing of layers
    b) Rearing of broilers
    c) Rearing of chicks up to the age of 8 weeks
    d) All of these
35. ..... Is used for growing the grower chicks from age of 8 weeks to 18 weeks.
    a) Brooder house
    b) Layer house
    c) Broiler house
    d) Grower house
36. Selection criteria for poultry house is ..... .
    a) Poultry farm should have a regular transport service
    b) Water and electricity facility must be available
    c) Poultry house should dace south or north
    d) All of these
37. The feeding equipment used in deep litter system are.................. .
    a) Tube feeders
    b) Pan feeders
    c) Long food trough
    d) All of these
38. Advantage of the deep litter system is ..... .
    a) Birds are more comfortable in deep litter system
    b) Used litter is a very good manure
    c) Eggs and flesh production are high
    d) All of these

39. Cage system is ....
    a) An Intensive method
    b) Made up of wire-mesh
    c) Constructed in poultry house
    d) All of these
40. Advantage of cage system is ....
    a) It needs limited space
    b) Cage birds consume less feed
    c) Cage system needs less labour
    d) All of these
41. Disadvantages of cage system are ....
    a) Bad smelling
    b) Cracked eggs
    c) Birds consume more feed
    d) Both a & b
42. Marker's disease vaccine is given to chicks ....
    a) On first day
    b) Subcutaneously
    c) Introcular
    d) Both a & b
43. Ranikhet disease vaccine-'F' strain is given to chicks ..... .
    a) On 5th day
    b) Intranasal and intraocular
    c) Both a & b
    d) On 8th week
44. Gumboro disease vaccine is given in ....
    a) 1st week
    b) 3rd week
    c) Intraocular
    d) Both b & c
45. Ranikhet disease vaccine (RDVK) is administered subcutaneously in ....
    a) 8th week
    b) 3rd week
    c) 1st day
    d) 3 hrs
46. Growers are ....
    a) The hens between the age group of 8th week to 18th week
    b) The hens between the age group of 3 to 4th week
    c) The hens between the age group of 5th to 8th week
    d) The hens between the age group of 18th to 21st week
47. The chicks need the following schedule of temperature
    a) First week — i) 70°F
    b) Second week — ii) 65°F
    c) Third week — iii) 80°F
    d) Fourth week — iv) 80°F

e) Fifth week v) 90°F

f) Sixth week vi) 85°F

g) Seventh week vii) 95°F

a) a=iii, b=vii, c=ii,d=i,e=iv,f=v,g=v

b) a=vii, b=v, c=vi,d=iii,e=iv,f=i, g=ii

c) a=vii, b=v, c=iii, d=vi, e=iv, f=i, g=ii

d) a=vii, b=v, c=vi, d=iii,e=ii, f=i, g=iv

48. ….. Is calculated by dividing the total feed consumed by the total weight of the bird.

a) Forage ratio b) Feed conversion ratio

c) Feeding intensity d) Both b & c

49. The feed conversion ratio for an ideal bird is ….. .

a) 8 b) 6

c) 2 d) 9

50. Ranikhet disease is also known as ….. .

a) Fowl fox b) New castle disease

c) Coryza d) All of these

51. Ranikhet disease is a ….. .

a) Protozoan disease b) Fungal disease

c) Viral disease d) Worm disease

52. New Castle Disease affects ….. of poultry birds.

a) Respiratory system b) Nervous system

c) Urinary system d) Both a & b

53. Ranikhet disease is caused by ….. .

a) Eimera tenella b) Myxovirus multiforme

c) Haemophilus gallinarum d) Costia necatrix

54. Common symptoms of Ranikhet disease in poultry birds are ….. .

a) Coughing and gasping b) Paralysis of legs and wings

c) Rattling of the wind pipe d) All of these

55. Fowl fox is a ….. .

a) Viral disease b) Fungal disease

c) Bacterial disease d) Protozoan disease

56. Fowl fox is characterized by …. . .
    a) Scales and cankers on the comb
    b) Scales and cankers on the mouth
    c) White spots on cloacal region
    d) Both a & b

57. Fowl fox is caused by …. . .
    a) Eimeria tenella
    b) Haemophilus gallinarum
    c) Borreliota avium
    d) Myxovirus multiforme

58. Treatment for fowl fox is …. . .
    a) High dose of vitamin A
    b) Antimony tartarate-200
    c) Quinine hydrochloride
    d) Both a & b

59. …. . is a bacterial disease characterized by 'Cold in poultry birds'.
    a) Fowl fox
    b) Coccidiosis
    c) New castle disease
    d) Coryza

60. …. . is caused by the bacterium Haemophilus gallinarum.
    a) Coryza
    b) Fowl fox
    c) Ranikhet disease
    d) New castle disease

61. Coryza is treated by …. . .
    a) Antimony tartatrate-200
    b) Vitamin A
    c) Vitamin D
    d) Tetracycline

62. Coccidiosis is caused by …. . .
    a) Eimeria
    b) Geometra
    c) Costia
    d) Argulus

63. The symptoms of coccidiosis include…. . .
    a) Bloody droppings
    b) Diarrhea
    c) Loss of weight
    d) All of these

64. By which drug coccidiosis is treated?
    a) Sulphadimidine
    b) Sulphaquinoxaline
    c) Nitrofurazone
    d) All of these

65. Polyneuritis is a disorder of …. . .
    a) Respiratory system
    b) Circulatory system
    c) Nervous system
    d) Digestive system

66. ….. is caused due to the deficiency of Vitamin.
    a) Polyneuritis b) Coccidiosis
    c) New castle disease d) Fowl fox
67. Important step for preparation of shed before brooding is …… .
    a) To spray an insecticide just after the old birds have been removed
    b) To remove all non-stationary equipment and residual feed troughs and bins
    c) To eliminate all rodents and wild birds if any
    d) All of these
68. ….. is obtained by grinding and mixing of the raw materials.
    a) Mash feed b) Pellet feed
    c) Live feed d) Crumbles
69. ….. is prepared by exposing the mash feed to heat treatment under pressure.
    a) Crumbles b) Mash-raw feed
    c) Pellet feed d) Both b & c
70. ….. Preparation is little expensive compared to mash and pellets.
    a) Live feed b) Crumbles
    c) Artificial feed d) Natural food
71. Fowl cholera is caused by …… .
    a) Pasterulla multocida b) Sallmonella spp
    c) Eimeria tenella d) None of these
72. Fowl cholera is treated by …… .
    a) Enrofloxacin in water b) Injection of gentamycin
    c) Both a & b d) Aeroflavin injection
73. The standard hatchery operations includes …… .
    a) Hatching eggs receiving b) Fumigation
    c) Transfer of hatches d) All of these
74. The wastes generated in poultry farm are …… .
    a) Litter waste b) Dead birds
    c) Hatchery waste d) All of these
75. ….. Constitutes around 65-70% of the broiler market.
    a) The cob breed b) The Ters breed
    c) The charm breed d) Both a & b

76. India's per capita consumption of chicken meat is ....
    a) 2 kg/person/year
    b) 7 kg/person/year
    c) 18.95 kg/person/year
    d) 12 kg/person/year
77. The chicken meat contains ....
    a) Choline
    b) Vitamin B12
    c) Iron & Zink
    d) All of these
78. 'Aseel' is a ....
    a) Chinese breed
    b) Indigenous breed
    c) American breed
    d) Australian breed
79. Which of the following is a desi (Indigenous) breed?
    a) Aseel
    b) Ghagus
    c) Bustra
    d) All of these
80. ..... . is one of the most popular breeds of USA.
    a) Malay
    b) Chittagong
    c) Polymouth rock
    d) Fasciatus
81. Poultry manure is obtained from ....
    a) Dropping of poultry birds
    b) Feathers of poultry birds
    c) Bones of poultry bords
    d) All of these
82. Poultry manure contains ....
    a) Nitrogen (2%)
    b) Phosphoric acid (1.25%)
    c) Potash (0.75%)
    d) All of these
83. The number of eggs laid by an hen on consecutive days without break is called as ....
    a) Cluster
    b) Capon
    c) Pullet
    d) Clutch
84. ..... . is a young male bird whose testicles are removed usually below 8-10 months of age.
    a) Clutch
    b) Capon
    c) Pullet
    d) Hatchability
85. ..... . is a young female chicken of 9-20 weeks of age that has not yet started to lay eggs.
    a) Pullet
    b) Mullet
    c) Capon
    d) Cluster

86. Which disease is known as 'Fowl plague'?
    a) Manmodi b) Coccidiosis
    c) Bird flu d) Swine flu
87. Which chicken breed lays the most eggs in India?
    a) Optigton b) Leesbree
    c) Polymouth Rock d) All of these
88. Which of the following is not included in poultry farming?
    a) Duck b) Chicken
    c) Cow d) Turkey
89. Head quarter of Central Poultry Development Organization and Training Institute is located at ...
    a) Nagpur b) Mumbai
    c) New Delhi d) Bangalore
90. The Directorate of Poultry Research functions under ..... .
    a) ICAR b) NFDB
    c) CIFE d) APEDA
91. Central Avian Research is situated at ....
    a) Bhopal b) Izatnagar
    c) Tatanagar d) Raurkela
92. ..... was developed by crossing superior male and female strains of white leghorn.
    a) CARI Priya b) CARI Sonali
    c) CARI Dembendra d) CARI Plumage
93. Match the pairs

a) CARI Priya	i) Commercial egg laying chicken
b) CARI Debendra	ii) White broiler
c) CARI Sonali	iii) Dual purpose chicken
d) CARIBRO Vishal	iv) Commercial white egg laying chicken
e) CARIBODhanaraja	v) Suitable for hot and dry region
f) CARIBO Mrityunjay	vi) Coloured broiler

a) a=ii, b=iv, c=i, d=iii, e=vi, f=v
b) a=iv,b=iii, c=ii, d=i, e=vi, f=v
c) a=iv, b=iii, c=i, d=ii, e=vi, f=v
d) a=iv, b=iii, c=v, d=ii, e=i, f=vi

94. ….. is a cross of Indian native breed Aseel with CARI Red.
a) CARI Nirbheek
b) CARI Shyama
c) Upkari
d) Both a & b

95. ….. is a cross of Kadaknath breed of Indian native chicken with CARI Red.
a) CARI Nirbheek
b) Upkari
c) CARI Shyama
d) Hitcari

96. ….. are developed by crossing the Indian native naked neck with exotic breed CARI Red.
a) CARI Shyama
b) Hitcari
c) Upkari
d) Tridisal

97. ….. is a variety of Guncari bird.
a) Kadambari
b) Swetambari
c) Chitambari
d) All of these

98. …. . is a dual purpose chicken variety to cater to the needs of rural poultry keepers of Rajasthan
a) Jharsin
b) Busra
c) Pratapdhan
d) Chittagong

99. ….. is a dual purpose variety of free range farming in Assam.
a) Jharsin
b) Kamrupa
c) Chittagong
d) Busra

100. ….. said to be the home of CARI Nirbheek (Aseel cross).
a) West Bengal
b) Maharashtra
c) Assam
d) Andhra Pradesh

101. ….. of our country accounts for 4.11% of the GDP.
a) Aquaculture sector
b) Agriculture sector
c) Livestock sector
d) Both a & b

102. ….. of our country accounts for 25.6% of the overall agricultural GDP.
a) Livestock sector
b) Industrial sector
c) Fisheries sector
d) Animal husbandry

103. The common diseases of poultry include …… .
a) Botolism
b) Fowl typhoid
c) Fowl cholera
d) All of these

104. ….. . constitute 4-9% of live weight of poultry bird.
a) Legs
b) Feathers
c) Droppings
d) Eggs

105. Which of the following is layer bird?
a) White Leghorn
b) Black Leghorn
c) Both a & b
d) Coryza

106. ….. . is an Australian dual purpose bird.
a) White leghorn
b) Australorp
c) Plymouth Rock
d) Optigton

107. The broiler chicks are vaccinated to control of …… .
a) Marker's disease
b) Ranikhet disease
c) Gumboro disease
d) All of these

108. …. are twisted in opposite directions and serve to keep the yolk centered.
a) Chalaza
b) Outer shell membrane
c) Blastoderm
d) Air cell

109. …. rests between the outer and inner membranes at the egg's larger end.
a) Chalaza
b) Cuticle
c) Air cell
d) White yolk

110. What does the chick use the yolk for?
a) To get water
b) To get nutrients
c) To get gametes
d) To get air inside

111. An embryo of chick with 4 pairs of somites is …… .
a) 20 hours old
b) 80 hours old
c) 24 hours old
d) 6 hours old

112. 18 hours embryo of chick contains …… .
a) 3 pairs of somites
b) Notochord
c) Primitive streak
d) All of these

113. 33 hours embryo of chick shows ….
a) 12-13 somites are formed
b) Large eye vesicles
c) 4 mm length
d) All of these

114. In the chick embryo of 96 hours of incubation ..... .
   a) The entire body has been turned through 90 degree
   b) The embryo lies with its left side on the yolk
   c) Optic cup shows the more developed lens
   d) All of these

115. The egg of hen is ..... .
   a) Macrolecithal b) Microlecithal
   c) Chemolecithal d) Both b & c

116. The allantois serves which functions for the embryo?
   a) The allantois serves as an embryonic urinary bladder
   b) It serves as air bladder
   c) It serves as respiratory organ
   d) None of these

117. The chick emerges after ..... days of incubation at 103°F.
   a) 10 days b) 12 days
   c) 21 days d) 38 days

118. Cleavage in chick is ..... .
   a) Discoidal b) Meroblastic
   c) Mesoblastic d) Both a & b

119. India ranks ..... in the world in egg production.
   a) 2nd b) 5th
   c) 3rd d) 1st

120. India ranks ..... in chicken meat production.
   a) 1st b) 3rd
   c) 2nd d) 5th

121. ..... refers to the period related to the production of eggs.
   a) Blue revolution b) Pink revolution
   c) Red revolution d) Silver revolution

122. Who is known as 'Father of the poultry industry in India'?
   a) Prof. K.H. Alikunhi b) Dr. B.V. Rao
   c) Dr. V.G. Jhingran d) Dr. M.R.Sinha

123. In year 2022 the total egg production in India was …. . .

a) 129.60 billion numbers
b) 1300 billion numbers
c) 30 billion numbers
d) 50 billion numbers

124. In year 2022, the per capita availability of egg in India was …. . .

a) 95 eggs/annum
b) 180 eggs/annum
c) 48 eggs/annum
d) 120 eggs/annum

125. Who is the biggest egg producer in India?

a) Bihar
b) Andhra Pradesh
c) Maharashtra
d) Karnataka

126. The 'Basic Animal Husbandry Statistics' is prepared by …. . .

a) Ministry of Fisheries, Animal Husbandry & Dairying
b) Indian Council of Agricultural Research
c) Poultry Research, India
d) None of these

127. Which season is the best to raise chickens?

a) Winter
b) Rainy
c) Summer
d) All of these

128. The disease caused by a tick is known as…. .

a) Spirochaetosis
b) White spot disease
c) Coccidiosis
d) None of these

129. The Indian Veterinary Research Institute is located at…. .

a) Patna (Bihar)
b) Pune (Maharashtra)
c) Chennai (Tamil Nadu)
d) Izatnagar (Uttar Pradesh)

130. National Institute of High Security Animal Diseases is located at…. .

a) Bhopal
b) Nagpur
c) Hyderabad
d) Delhi

Answer Keys

1	a)	Poultry keeping
2	d)	All of these
3	b)	Eggs
4	d)	All of these
5	a)	Flesh
6	a)	Table breeds

7	c)	Both a & b
8	c)	General purpose birds
9	c)	Rhode Island Reds
10	d)	All of these
11	b)	Protein
12	d)	All of these
13	d)	All of these
14	d)	All of these
15	d)	All of these
16	d)	All of these
17	b)	Italy
18	d)	Both a & b
19	c)	Australorp
20	d)	All of these
21	b)	Vent sexing
22	b)	Female
23	d)	Japanese
24	a)	Dr. Kizoshi Masui (1927)
25	a)	Red
26	c)	Both a & b
27	d)	Both b & c
28	c)	Both a & b
29	d)	All of these
30	a)	Poultry house
31	d)	Deep litter house
32	d)	All of these
33	a)	Rearing of layers
34	b)	Rearing of broilers
35	d)	Grower house
36	d)	All of these
37	d)	All of these
38	d)	All of these
39	d)	All of these
40	d)	All of these
41	d)	Both a & b
42	d)	Both a & b
43	c)	Both a & b
44	d)	Both b & c
45	a)	8th week

46	a)	The hens between the age group of 8th week to 18th week
47	b)	a=vii, b=v, c=vi,d=iii,e=iv,f=I, g=ii
48	b)	Feed conversion ratio
49	c)	2
50	b)	New castle disease
51	c)	Viral disease
52	d)	Both a & b
53	b)	Myxovirus multiforme
54	d)	All of these
55	a)	Viral disease
56	d)	Both a & b
57	c)	Borreliota avium
58	d)	Both a & b
59	d)	Coryza
60	a)	Coryza
61	d)	Tetracycline
62	a)	Eimeria
63	d)	All of these
64	d)	All of these
65	c)	Nervous system
66	a)	Polyneuritis
67	d)	All of these
68	a)	Mash feed
69	c)	Pellet feed
70	b)	Crumbles
71	a)	Pasterulla multocida
72	c)	Both a & b
73	d)	All of these
74	d)	All of these
75	a)	The cob breed
76	b)	7kkg/person/year
77	d)	All of these
78	b)	Indigenous breed
79	d)	All of these
80	c)	Polymouth rock
81	d)	All of these
82	d)	All of these
83	d)	Clutch
84	b)	Capon

85	a)	Pullet
86	c)	Bird flu
87	c)	Polymouth Rock
88	c)	Cow
89	d)	Bangalore
90	a)	ICAR
91	b)	Izatnagar
92	a)	CARI Priya
93	c)	a=iv, b=iii, c=i, d=ii, e=vi, f=v
94	a)	CARI Nirbheek
95	c)	CARI Shyama
96	b)	Hitcari
97	d)	All of these
98	c)	Pratapdhan
99	b)	Kamrupa
100	d)	Andhra Pradesh
101	c)	Livestock sector
102	a)	Livestock sector
103	d)	All of these
104	b)	Feathers
105	c)	Both a & b
106	b)	Australorp
107	d)	All of these
108	a)	Chalaza
109	c)	Air cell
110	b)	To get nutrients
111	c)	24 hours old
112	d)	All of these
113	d)	All of these
114	d)	All of these
115	a)	Macrolecithal
116	a)	The allantois serves as an embryonic urinary bladder
117	c)	21 days
118	d)	Both a & b
119	c)	3rd
120	d)	5th
121	d)	Silver revolution
122	b)	Dr. B.V. Rao
123	a)	129.60 billion numbers

124	a)	95 eggs/annum
125	b)	Andhra Pradesh
126	a)	Ministry of Fisheries, Animal Husbandry & Dairying
127	b)	Rainy
128	a)	Spirochaetosis
129	d)	Izatnagar (Uttar Pradesh)
130	a)	Bhopal

Table 1: Agriculture Revolutions in India

Agricultural Revolution	Product
Green Revolution	Foodgrains Production
Blue Revolution	Fish Production
Yellow Revolution	Oil seed Production
Pink Revolution	Onion Production/Prawn Production
Red Revolution	Meat Production/Tomato Production
Brown Revolution	Cocoa Production
Golden Revolution	Fruit Production/Honey Production
Golden Fibre Revolution	Jute Production
Silver Revolution	Egg Production
Silver fiber Revolution	Cotton Production
Black Revolution	Petroleum
White Revolution	Milk Production

Table 2: Expansions of Acronyms

1. CBRTI= Centre Bee Research and Training Institute
2. HACCP= Hazard Analysis and Critical Control Point
3. FSSAI= Food Safety and Standards Authority of India
4. APEDA=Agricultural and Processed Food Products Export Development Authority
5. BIS= Bureau of Indian Standards
6. PVCF= Poultry Venture Capital Fund
7. EDEG= Entrepreneurship Development and Employment Generator
8. NABARD= National Bank for Agricultural and Rural Development
9. IPPP= Innovative Poultry Productivity Project
10. ASCAD= Assistance to States for Control of Animal Diseases
11. CARI= Central Avian Research Institute

12. RDDL= Regional Disease Diagnostic Laboratory
13. NIHSAD= National Institute of High Security Animal Diseases
14. HPAI= Highly Pathogenic Avian Influenza
15. NMSA= National Mission for Sustainable Agriculture
16. NFSM= National Food Security Mission
17. MIDH= Mission Integrated Development of Horticulture
18. RKVY= Remunerative Approaches for Agriculture and Allied Sector Rejuvenation
19. NCOF= National Centre of Organic Farming

9

Apiculture

1. 'Apis' means …… .
   a) Waxf
   b) Bee
   c) Honey
   d) All of these
2. Apiculture is also known as …… .
   a) Bee-keeping
   b) Wax preparation
   c) Both a & b
   d) None of these
3. Importance of bee-keeping …… .
   a) It provides honey
   b) It provides bee wax
   c) Honey bees are excellent pollinating agents, thus increasing agricultural
   d) All of these
4. Species of honey bee is …… .
   a) Apis dorsata
   b) Apis indica
   c) Apis florae
   d) All of these
5. …. . is ferocious, stings severely causes fever and sometimes even death,
   a) Apis indica
   b) Apis dorsata
   c) Apis mellifera
   d) Apis florae
6. ….. is commonly known as Little bee.
   a) Apis indica
   b) Apis dorsata
   c) Apis florea
   d) None of these
7. Match the pairs
   a) Apis dorsata — i) The Indian bee
   b) Apis indica — ii) The European bee
   c) Apis florae — iii) The little bee

d) Apis mellifera iv) The rock bee

a) -a)=ii, b)=i, c)=iv, d)=iii
b) -a)=iii, b)=ii, c)=iv, d)=i
c) -a)=iv, b)=i, c)=iii, d)=ii
d) -a)=iv, b)=iii, c)=i, d)=ii

8. Name the species of honeybee that can be domesticated.
a) Apis indica
b) Apis mellifera
c) Both a & b
d) Apis dorsata

9. A honeybee colony has ….. castes.
a) Two
b) Three
c) Four
d) Five

10. ….. is single and functional female.
a) Queen bee
b) Worker
c) Drone
d) Both a & b

11. Drones are ….. .
a) Female bees
b) Male bees
c) Workers
d) All of these

12. Worker bees are ….. .
a) Smaller than the Queen
b) With strong wings to fly
c) With large and efficient proboscis
d) All of these

13. ….. have a life span of about 35 days.
a) The worker
b) Queen bee
c) Drones
d) Both b & c

14. The worker's duty from day 1-14 is ….. .
a) Guard duties at entrance to the hive
b) Foraging
c) Collecting the food
d) All of these

15. The worker's duty from day 14-20 is ….. .
a) Guard duties at entrance of the larvae the hive
b) Collecting the food
c) Cleaning the hive
d) Feeding

16. The worker's duty from day 21-35 is …… .
   a) Feeding the larvae
   b) Hive cleaning
   c) Foraging
   d) All of these
17. ….. . visit flower to flower, collect nectar and pollen and return to their own nest.
   a) The workers
   b) Queen bee
   c) Drones
   d) Both a & b
18. ….. . are the males produced from unfertilized eggs.
   a) The workers
   b) Drones
   c) Both a & b
   d) None of these
19. Drones can live up to …… .
   a) 10 days
   b) 15 days
   c) 60 days
   d) 130 days
20. At the age of ….. . the drones perform mating fight chasing the virgin Queen in the air.
   a) 30-40 days
   b) 5-9 days
   c) 14-18 days
   d) 50-60 days
21. Indigenous method of bee keeping includes …… .
   a) Wall or fixed types of hives in rectangular spaces in the walls with a small hole
   b) Movable types of hives in wooden boxes or earthen pitchers
   c) Both a & b
   d) Series of square or oblong boxes without tops or bottoms, set one above the other.
22. The plants which yield nectar and pollen are collectively termed as….. .
   a) Bee pasturage
   b) Swarm
   c) Skep
   d) Supers
23. ….. . is a food material for the bees and their larvae.
   a) Wax
   b) Honey
   c) Swarm
   d) Skep
24. Honey contains …… .
   a) Water (13-20%)
   b) Fructose (40-50%)
   c) Glucose (2-3%)
   d) All of these

25. …. . is a disaccharide which is hydrolysed by the salivary amylase to produce monosaccharides.

a) Honey stomach b) Nectar

c) Honey d) Both a & b

26. Nectar is sucked from flowers, mixed with saliva and swallowed into a special gut region called …. . .

a) Proboscis b) Bee pasturage

c) Honey stomach d) Honey sting

27. Uses of honey include …. . .

a) Nutritious food

b) Used as carrier in ayurvedic medicines

c) Used in religious ceremonies

d) All of these

28. Uses of beewax include …. . .

a) Making of candles

b) Preparation of varnishes and paints

c) Making pharmaceutical preparations

d) All of these

29. …. . is located on the underside of the last four abdominal segments (4th -7th) of the worker bee.

a) Proboscis b) Sting

c) Wax gland d) Honey stomach

30. Beekeeping is also known as …. . .

a) Apiculture b) Aquaculture

c) Pollination d) Propolis

31. …. . is the father of Apiculture and Apiology.

a) Moses Quinby b) M.Langstroth

c) Johann Zierzon d) T.Treast

32. …. . Is the father of commercial Beekeeping.

a) Langstroth b) Moses Quinby

c) J. Dzierzon d) Linnaeus

33. …. . are the modern hives.

a) Top bar hives b) Newton hive

c) Langstroth hive d) All of these

34. Which method is practiced in Kashmir & Karnataka ?
    a) Wall hive b) Pot hive
    c) Log hive d) Bamboo hive
35. Which is globally modern beekeeping method ?
    a) Top bar hive b) Langstroth hive
    c) Newton hive d) Both a & b
36. ….. is a wooden box used by Indians for rearing honey bees.
    a) Top bar hives b) Newton hive
    c) Langstroth hive d) Both a & b
37. ….. is a board used to keep the Queen inside the brood chamber and to prevent it to enter the super chamber.
    a) Queen excluder b) Hung
    c) Newton hive d) Langstroth hive
38. ….. is an open wooden box containing honey comb frames.
    a) Queen excluder b) Super chamber
    c) Top cover d) Hung
39. Advantages of Newton Hive are …...
    a) It facilitates easy harvesting
    b) The combs can be reused
    c) Bees are protected from enemies and weather
    d) All of these
40. Advantages of modern Beekeeping…...
    a) Artificial feeding is possible
    b) Swarming is controlled
    c) Bee is managed for pollination
    d) All of these
41. ….. is the bee yard where the hives are installed.
    a) Apiary b) Grafting
    c) Brood nest d) None of these
42. ….. is used to calm honey bees.
    a) A smoker b) Robbing
    c) Acariosis d) Varrosis

43. Drones are trapped by …… .
   a) Robbing b) Varrosis
   c) Drone excluder d) All of these
44. …… is kept between brood chamber and super chamber.
   a) Queen excluder b) Drone excluder
   c) Varrosis d) Robbing
45. Bee wax is the …… .
   a) Ester of fatty acids b) Sticky and insoluble in water
   c) Bee sweat d) All of these
46. …… is secreted by worker bees to construct the comb.
   a) Bees-wax b) Honey
   c) Nectar d) None of these
47. Number of wax glands in worker bees are …… .
   a) Five b) Six
   c) Seven d) Eight
48. The ambient temperature for wax secretion in the hive to be …… .
   a) 7-10°C b) 15-20°C
   c) 33-36°C d) 40-42°C
49. By which method bee wax is extracted?
   a) Hot water extraction b) Wax press method
   c) Solar wax method d) All of these
50. Bees wax is used …… .
   a) To treat lupus b) To prepare lotions
   c) In sculptures d) All of these
51. Bee venom is known as …… .
   a) Apitoxin b) Apidae
   c) Bee sweat d) Both b & c
52. Bee venom is …… .
   a) An anticoagulant b) Soluble in water
   c) Insoluble in alcohol d) All of these
53. The main components of venom are…… .
   a) Proteins b) Enzymes
   c) Peptides d) All of these

54. Bee venom is known as ....
    a) Curative toxin
    b) Therapeutic agent
    c) Both a & b
    d) Scar tissue
55. Bee venom is used to ....
    a) Cure hypertension
    b) Cure asthma
    c) Cure headache
    d) All of these
56. A swarm consists of ....
    a) A Queen
    b) Workers
    c) Drones
    d) All of these
57. Reason for swarming is ....
    a) Insufficient amount of Queen pheromone
    b) Congestion
    c) Missing Queen
    d) All of these
58. Symptoms of swarming includes ....
    a) An increased rate of egg laying
    b) Construction of drone cells on the two sides of comband drone rearing
    c) Construction of swarm cells on the lower edge of the comb
    d) All of these
59. Advantages of swarming ....
    a) Swarming produces new colonies
    b) A new colony can be prepared
    c) Requeening can be done
    d) All of these
60. Swarming is controlled by ....
    a) The demorec method
    b) Clipping of wings
    c) Artificial swarming
    d) All of these
61. Artificial swarming has the following advantages...... .
    a) Prevents natural swarming
    b) Helps to produce new colonies
    c) Helps in requeening
    d) All of these

62. Hopkins method is used for ..... .

a) Queen rearing
b) Swarming
c) Wax extraction
d) Honey extraction

63. ..... is a non-grafting method of queen rearing.

a) Hopkins method
b) Miller method
c) Both a & b
d) None of these

64. Queen rearing by Doolittle method is a ..... .

a) Non-grafting method
b) Grafting method
c) Both a & b
d) None of these

65. ..... Is the transfer of worker larvae of 24hrs old into artificial queen caps.

a) Non-grafting
b) Grafting
c) Royal jelly
d) Capping

66. In..... queen cells are raised on a prepared strip of cells.

a) Doolittle method
b) Alley method
c) Hopkins method
d) All of these

67. Queen rearing by Doolittle method requires ..... .

a) A good breeder queen
b) A cell builder colony
c) Larvae of the proper age
d) All of these

68. The breeder queen must be ..... .

a) Strong
b) Vigorous
c) Healthy
d) All of these

69. The queenless colony selected for raising the queen cells is known as ..... .

a) Cell builder colony
b) Cell starter
c) Cell cup
d) Cell bar

70. ..... Is used to transfer larva from cells into grafting cell cups for queen rearing.

a) Bar frame
b) Grafting tool
c) Starter colony
d) Queen cup

71. Nectare honey is ..... .

a) Acidic
b) Alkaline
c) Neutral
d) Turns acidic after few days

72. For how long does a worker bee live in the summer ?
    a) For 15 weeks   b) For 2 weeks
    c) For 2 months   d) For 6 months
73. The development of a male bee (dron e) takes how long?
    a) 15 days   b) 24 days
    c) 07 days   d) 5 days
74. Which of the following hormone is secreted by queen of honey bees?
    a) Copulin   b) Antiqueen Pheromone
    c) Trail pheromone   d) Bombyx
75. Which of the following bees do the tail-wagging dance?
    a) Nurse bees   b) Builder bees
    c) Sanitary bees   d) Scout bees
76. Which of the following is not a useful product of Apiculture?
    a) Bee venom   b) Bees wax
    c) Jelly   d) Honey
77. Which of the following does not attack honey bee?
    a) Wasps   b) Houseflies
    c) Mites   d) Black ants
78. Globally, for honey bee farm European Bee is preferred because of ….. .
    a) It has a gentle temper   b) It is less swarming
    c) It is a good honey gatherer   d) All of these
79. Queen's pheromone glands are located in ….. .
    a) Mandible   b) Proboscis
    c) Pollen basket   d) Antennae
80. ….. is located in the lower part of the young worker's abdomen.
    a) Pheromone gland   b) Hypophyseal gland
    c) Wax gland   d) Both a & b
81. Isopental acetate is produced by ….. .
    a) Scant glands   b) Hypophyseal gland
    c) Royal jelly   d) Wings
82. Queen is ….. .
    a) Fertile and functional female   b) Sterile female
    c) Male   d) Both a & b

83. Worker is ..... .

a) Fertile female
b) Sterile female
c) Male
d) Sterile male

84. Queen and worker develop from ..... .

a) Unfertilized egg
b) Fertilized egg
c) Ambulacral stage
d) None of these

85. Drone develops from ..... .

a) Fertilized egg
b) Pedicillariae
c) Unfertilized egg
d) Both a & b

86. Central Bee Research and Training Institute is located at..... .

a) Lucknow
b) Bhopal
c) Mumbai
d) Pune

87. Central Bee Research and Training Institute was established in ..... .

a) 1952
b) 1962
c) 1968
d) 1973

88. Objective of Central Bee Research and Training Institute is ..... .

a) To understand comprehensive training and research activities
b) To assess bee keeping impact on ecology, agriculture and horticulture
c) To create awareness on bee hive products
d) All of these

89. In India Sweet Revolution is known as ..... .

a) Mithi kranti
b) Honey mission
c) Both a & b
d) None of these

90. National Beekeeping and Honey Mission was launched in ..... .

a) 2020
b) 2017
c) 2012
d) 2014

91. Objectives of National Beekeeping and Honey Mission is..... .

a) Overall promotion and development of scientific beekeeping
b) To achieve the goal of 'Sweet Revolution'
c) Setting up of integrated Beekeeping Development Centres
d) All of these

92. 'World Bee Day' is celebrated on ..... every year.

a) 20th May
b) 2nd September
c) 12th January
d) 4th March

93. In ..... United Nations General Assembly unanimously proclaimed 20th May as 'World Bee Day'.

a) 2015
b) 2017
c) 2018
d) 2020

94. ..... is the first observance of World Bee Day.

a) 20th May 2018
b) 20th May 2020
c) 2nd September 2018
d) 4th March 2018

95. In year 2020-21, honey production in India was ..... .

a) 94500MTs
b) 1,05,000MTs
c) 50,000MTs
d) 1,25,000MTs

96. Which is the highest honey-producing state in India?

a) Maharashtra
b) Karnataka
c) Punjab
d) West Bengal

97. Which is the highest honey producing country in the world?

a) India
b) Russia
c) Nepal
d) China

98. Indian Government launched 'Sweet Revolution' in ..... .

a) 2017
b) 2015
c) 2012
d) 2014

99. Honey bees are the members of order ..... .

a) Pterygota
b) Hymenoptera
c) Insects
d) Apida

100. ..... Indicates that the sources of food is near the hive.

a) Round dance of hive
b) Wag tail dance
c) Swarming
d) Both a & b

101. If the source of food is more than 100 meters away, the informant bee performs ..... .

a) Round dance
b) Wag-tail dance
c) Pupae dance
d) Swarming

102. A place where bees are kept is known as ..... .
    - a) Apiary
    - b) Dorsata
    - c) Cocoon
    - d) Royal jelly

103. ..... .causes damage to honey bee colonies and to bee products in tropical and sub-tropical Asia.
    - a) Multifilis
    - b) The greater wax moth
    - c) Galleria
    - d) Apiary

104. The scientific name of the lesser wax moth is ..... .
    - a) Achroia grisella
    - b) Galleria mellonella
    - c) Apis mellifera
    - d) Apis cerana

105. The scientific name of the Greater wax moth is ..... .
    - a) Galleria mellonella
    - b) Achroia girisella
    - c) Apis mellifera
    - d) Apis cerana

106. Varroa mite is a native parasite of ..... .
    - a) Apis cerana
    - b) Apis mellifera
    - c) Apis florae
    - d) Apis dorsata

107. ..... is used to kill mites in the sealed brood cells.
    - a) Formalin
    - b) Picric acid
    - c) Quinine hydrochloride
    - d) Formic acid

108. Scientific name of tracheal mite is ..... .
    - a) Acarapis woodi
    - b) Acarine sepias
    - c) Tropilaelaps spp.
    - d) Varroa destructor

109. American foul brood disease is caused by ..... .
    - a) Paenibacillus larvae
    - b) Brachionus larvae
    - c) Rotatoria spp.
    - d) Both a & b

110. European foul brood disease is caused by ..... .
    - a) Streptococcus faeccalis
    - b) Achromobacter Eurydice
    - c) Melissococcus plutonius
    - d) Paenibacillus larvae

111. In India European foulbrood disease was first reported in ..... .
    - a) 1978
    - b) 1970
    - c) 1979
    - d) 1989

112. Nosema disease is a ..... .
    - a) Bacterial disease
    - b) Fungal disease
    - c) Viral disease
    - d) Lice disease

113. ….. is effective for nosema control.

a) Heat treatment b) Fumigation

c) Both a & b d) Dry smoke treatment

114. Who is father of Beekeeping in India ?

a) Raverend Newton (Kerala) b) C. Anita

c) K.R. Roy d) A.Q. Khan

115. India is at ….. position in the world for honey production.

a) 2nd b) 1st

c) 5th d) 8th

116. Duty of a drone is …...

a) To fertilize the queen

b) To help in maintenance of hive temperature

c) Both a & b

d) To maintain colony cochesion

117. Duty of queen is …...

a) To lay eggs in a colony

b) Prevent swarming of colonies

c) Prevent absconding of colonies

d) All of these

118. Tail wagging dance is also known as …...

a) Wag tail dance b) Round dance

c) Whirling dance d) Rotational dance

119. Honey bees belong to which family?

a) Apidae b) Hymenoptera

c) Serridae d) Hypotremata

120. Example of stingless honey bee is…..

a) Melipona sp b) Trigona sp

c) Both a & b d) Apidata

121. Body of honey bee consists of…..

a) Head b) Thorax

c) Abdomen d) All of these

122. Mouth parts of worker bees are modified for ..... .
    a) Sucking b) Lapping
    c) Both a & b d) Chewing
123. ..... are imperfect females and their duty is to clean the hive.
    a) Workers b) Queens
    c) Both a & b d) Stinging queen
124. ..... is used for preventing bee stings on face and neck.
    a) Bee veil b) Hive tool
    c) Bee brush d) Excluder
125. The success of bee keeping depends on ..... .
    a) Good bee b) Good apiary site
    c) Proper management d) All of these
126. Bee flora is also known as ..... .
    a) Bee forage b) Bee pasture
    c) Both a & b d) Swarm
127. Bee flora should have ..... .
    a) Long flowering period
    b) High density of flowers per unit of the plants
    c) Good quality of nectar with high concentration of sugars
    d) All of these
128. Swarming occurs due to ..... .
    a) Overcrowding and lack of ventilation
    b) Presence of old queen
    c) Sudden honey flow
    d) All of these
129. Symptoms of bee poisoning includes ..... .
    a) Bees become paralytic
    b) Bees become irritated and sting heavily
    c) Brood chilling due to reduced bee population
    d) All of these
130. Which of the following is bee enemy?
    a) Braulacoeca b) Meropsorientalis
    c) Galleria d) All of these

Answer Keys

1	b)	Bee
2	a)	Bee-keeping
3	d)	All of these
4	d)	All of these
5	b)	Apis dorsata
6	c)	Apis florea
7	c)	a) =iv, b)=i, c) d)=ii
8	c)	Both a & b
9	b)	Three
10	a)	Queen bee
11	b)	Male bees
12	a)	Smaller than the Queen
13	a)	The worker
14	d)	All of these
15	a)	Guard duties at entrance of the larvae the hive
16	c)	Foraging
17	a)	The workers
18	b)	Drones
19	c)	60 days
20	c)	14-18 days
21	c)	Both a & b
22	a)	Bee pasturage
23	b)	Honey
24	d)	All of these
25	b)	Nectar
26	c)	Honey stomach
27	d)	All of these
28	d)	All of these
29	c)	Wax gland
30	a)	Apiculture
31	c)	Johann Zierzon
32	b)	Moses Quinby
33	d)	All of these
34	b)	Pot hive
35	b)	Langstroth hive
36	b)	Newton hive
37	a)	Queen excluder
38	b)	Super chamber

39	d)	All of these
40	d)	All of these
41	a)	Apiary
42	a)	A smoker
43	c)	Drone excluder
44	a)	Queen excluder
45	d)	All of these
46	a)	Bees-wax
47	d)	Eight
48	c)	33-36°C
49	d)	All of these
50	d)	All of these
51	a)	Apitoxin
52	d)	All of these
53	d)	All of these
54	c)	Both a & b
55	d)	All of these
56	d)	All of these
57	d)	All of these
58	d)	All of these
59	d)	All of these
60	d)	All of these
61	d)	All of these
62	a)	Queen rearing
63	c)	Both a & b
64	b)	Grafting method
65	b)	Grafting
66	b)	Alley method
67	d)	All of these
68	d)	All of these
69	a)	Cell builder colony
70	b)	Grafting tool
71	a)	Acidic
72	c)	For 2 months
73	b)	24 days
74	b)	Antiqueen Pheromone
75	d)	Scout bees
76	c)	Jelly
77	b)	Houseflies

78	d)	All of these
79	a)	Mandible
80	c)	Wax gland
81	a)	Scant glands
82	a)	Fertile and functional female
83	b)	Sterile female
84	b)	Fertilized egg
85	c)	Unfertilized egg
86	d)	Pune
87	b)	1962
88	d)	All of these
89	c)	Both a & b
90	a)	2020
91	d)	All of these
92	a)	20th May
93	b)	2017
94	a)	20th May 2018
95	d)	1,25,000MTs
96	c)	Punjab
97	d)	China
98	a)	2017
99	b)	Hymenoptera
100	b)	Wag tail dance
101	b)	Wag-tail dance
102	a)	Apiary
103	b)	The greater wax moth
104	a)	Achroia grisella
105	a)	Galleria mellonella
106	a)	Apis cerana
107	d)	Formic acid
108	a)	Acarapis woodi
109	a)	Paenibacillus larvae
110	c)	Melissococcus plutonius
111	b)	1970
112	b)	Fungal disease
113	b)	Fumigation
114	a)	Raverend Newton (Kerala)
115	d)	8th
116	c)	Both a & b

117	d)	All of these
118	a)	Wag tail dance
119	a)	Apidae
120	c)	Both a & b
121	d)	All of these
122	c)	Both a & b
123	a)	Workers
124	a)	Bee veil
125	d)	All of these
126	c)	Both a & b
127	d)	All of these
128	d)	All of these
129	d)	All of these
130	d)	All of these

10

Sericulture

1. ….. . involves rearing of silkworms.
   a) Lac culture
   b) Sericulture
   c) Apiculture
   d) Pisciculture
2. 'Silk Samagra' scheme was launched by ….. .
   a) Government of Karnataka
   b) Government of Tamil Nadu
   c) Government of Punjab
   d) Government of India
3. Match the pairs

a)	Mulberry silkworm	1.	Antheraea paphia
b)	Tasar silkworm	2.	Attacus ricinii
c)	Muga silkworm	3.	Antheraea pernyi
d)	Eri silkworm	4.	Attacus altas
e)	Oak silkworm	5.	Bombyx mori
f)	Giant silkworm	6.	Antheraea assama

   a) a=5, b=2,c=3,d=4,e=1, f=6
   b) a=5, b=1, c=6, d=2, e=3, f=4
   c) a=2, b=4, c=3, d=1, e=6, f=5
   d) a=2, b=3, c=2, d=1, e=4, f=6
4. 'Silk Samagra' scheme was launched for period of ….. .
   a) 2017-2020
   b) 2016-2019
   c) 2014-2017
   d) 2021-2024
5. The non-mulberry varieties of silks are known as ….. .
   a) Vanya silks
   b) Wild silks
   c) Both a & b
   d) Paithani silks
6. Match the pairs

a)	Andhra Pradesh	i)	Kollegal
b)	Bihar	ii)	Champa
c)	Gujrat	iii)	Narainpet

d) Karnataka | iv) Paithan
e) Chattisgarh | v) Bhagalpur
f) Maharashtra | vi) Kumbhakonam
g) Tamil Nadu | vii) Surat

a) a=i, b=iv, c=ii, d=v, e=iii, f=vi, g=vii
b) a=iii, b=ii, c=iv, d=i, e=v, f=vii, g=vi
c) a=iii, b=v, c=vii, d=i, e=ii, f=iv, g=vii
d) a=iii, b=iv, c=vi, d=v, e=vii, f=i, g=ii

7. Famous silk centre in West Bengal is …… .
   a) Birbhum
   b) Bishnupur
   c) Barrackpore
   d) Both a & b
8. …… . were the first locations to have automated silk reeling units.
   a) Hyderabad (Andhra Pradesh)
   b) Gobichetti palayam (Tamil Nadu)
   c) Both a & b
   d) Varanasi and Paithan
9. India imports raw silk for its industry from …… .
   a) China b) France
   c) Denmark d) Germany
10. The production of muga silk is mostly confined to …… .
   a) Maharashtra b) Bihar
   c) Orissa d) Assam
11. The major production of tassar raw silk is achieved from …… .
   a) Bihar and Orissa
   b) Madhya Pradesh & Andhra Pradesh
   c) West Bengal
   d) All of these
12. The eri raw silk production is achieved by …… .
   a) Assam & Bihar b) Meghalaya & Manipur
   c) Maharashtra & Uttar Pradesh d) Both a & b

13. 'National Sericulture Project' was launched in ..... .
    a) 1980
    b) 1989
    c) 1999
    d) 2016
14. 'National Sericulture Project' was launched in .... . Indian states.
    a) Three
    b) Four
    c) Five
    d) Six
15. The objectives of 'National Sericulture Project' were ..... .
    a) Introduction of mulberry sericulture in non-mulberry silk states
    b) Expansion of sericulture industry in mulberry silk producing states
    c) Generation of employment opportunities for an additional one million people
    d) All of these
16. Central Sericulture Research and Training Institute is located at....
    a) Berhampore
    b) Bangalore
    c) Paithan
    d) Varanasi
17. Central Silk Technological Research Institute is located at .... .
    a) Mumbai
    b) Berhampore
    c) Bangalore
    d) Chennai
18. Central Silk Board was constituted in ..... .
    a) 1959
    b) 1999
    c) 1948
    d) 1979
19. Function of Central Silk Board is ..... .
    a) Proper development of sericulture industry in country
    b) To promote and encourage scientific, technological and economic research for further advancement
    c) To develop healthy silkworm seed and ensure their proper distribution
    d) All of these
20. The Regional Tasar Research Station on Temperate Oak Tasar came into existence in ..... .
    a) 1972
    b) 1982
    c) 1992
    d) 2002

21. International Centre for Training and Research in Tropical Sericulture (ICTRITS) is located at which place?
    a) Bangalore
    b) Gandhinagar
    c) Mysore
    d) Anantpur
22. Central Silk Technological Research Institute was established in ..... .
    a) 1983
    b) 1989
    c) 2001
    d) 1979
23. Eri silk Research Station was established in ..... .
    a) Boko (Assam)
    b) Imphal (Manipur)
    c) Bhimtal (Uttar Pradesh)
    d) Mendipathor (Meghalaya)
24. Oak Tasar Silk Research Station is located at ..... .
    a) Batote (Jammu & Kashmir)
    b) Bhimtal (Uttar Pradesh
    c) Imphal (Manipur)
    d) All of these
25. Muga silk Research Station was established at ..... .
    a) Bhimtal (Uttar Pradesh)
    b) Manjra (Uttar Pradesh)
    c) Buko (Assam)
    d) Dhule (Maharashtra)
26. Native place of Mulberry silkworm is ..... .
    a) Brazil
    b) China
    c) Italy
    d) India
27. Bombyx mori is known as mulberry silkworm because ..... .
    a) Its natural food is mulberry leaf
    b) Its breeding takes place on mulberry leaf
    c) Both a & b
    d) Its origin is from Mulberry region of China
28. Tasar silkworm belongs to which family?
    a) Bombycidae
    b) Saturniidae
    c) Mulberridae
    d) Anabaatidae
29. The caterpillars of muga silkworm feed on .... .
    a) Castor plant
    b) Arandi Plant
    c) Machilus plant
    d) All of these
30. .... . Feeds on caster leaf.
    a) Eri silkworm
    b) Muga Silkworm
    c) Tasar silkworm
    d) Giant silkworm

31. The race of silkworm by which only one crop is taken in one year is called ..... .

a) Multi-voltine
b) Uni-voltine
c) Bi-voltine
d) Di-voltine

32. The race of silkworm by which more than two crops taken in a year is called ..... .

a) Multi-voltine
b) Uni-voltine
c) Bi-voltine
d) Bi-voltine

33. In mulberry silkworm fertilization is ..... .

a) External
b) Internal
c) On mulberry leaves
d) On land

34. ..... eggs are laid by the silkworm inhabiting in temperate regions.

a) Non-diapause
b) Diapause
c) Menopause
d) Polypause

35. Silkworms belonging to subtropical regions like India lay ..... .

a) Diapause eggs
b) Non-diapause eggs
c) Polypause eggs
d) Menopause eggs

36. Eggs of Mulberry silkworm after ten days of incubation hatch into a larvae called ..... .

a) Bombyx
b) Axol
c) Caterpillar
d) Nepa

37. The caterpillar is ..... .

a) Twelve segmented
b) Pale, yellow-white in colour
c) Both a & b
d) Plankton feeder

38. Weight of the full grown caterpillar varies from ..... .

a) 20-25gm
b) 10-15gm
c) 4-6gm
d) 25-35gm

39. .... . Is the white coloured bed of the pupa whose outer threads are irregular while the inner threads are regular ..... .

a) Cocoon
b) Imago
c) Machana
d) Morus

40. Sericulture requires ..... .

a) Machana
b) Rearing trays
c) Chopping knife
d) All of these

41. Mulberry tree belongs to which family?
    a) Moraceae b) Saturniidae
    c) Bombycidae d) Cyanophyceae
42. ….. is locally known as 'Toot' or 'Tooti'.
    a) Morus indica b) Morus alba
    c) Morus pyrix d) None of these
43. ….. is locally known as 'Shabtoot'.
    a) Morus alba b) Morus pyrix
    c) Morus indica d) Morus zebi
44. Plantation of Mulberry is done by …...
    a) Seed sowing b) Cutting
    c) Both a & b d) None of these
45. The quality of cocoon depends on …...
    a) Raw silk yield b) Filament length
    c) Splitting d) All of these
46. Post-cocoon processing includes …...
    a) Stifling b) Reeling
    c) Both a & b d) None of these
47. The process of killing the cocoons is termed as …...
    a) Reeling b) Stifling
    c) Spinning d) Laying
48. The killing of the cocoon in boiling water helps in …...
    a) Softening the adhesion of the silk threads
    b) Loosening of the outer threads to separate freely
    c) Facilitating the unbinding of silk threads
    d) All of these
49. The process of removing the threads from the killed cocoon is called as …...
    a) Stifling b) Reeling
    c) Slanting d) Both b & c
50. Raw silk is also known as …...
    a) Reeled silk b) Raw silk
    c) Both a & b d) Slant

51. ….. is the fibrous protein secreted by silkworm.
 a) Silk b) Yarn
 c) Nepa d) Chorion

52. Cultivation of Mulberry plants for the production of mulberry leaves is known as …. . .
 a) Morus culture b) Serrata culture
 c) Moriculture d) Sericulture

53. Varieties of Mulberry plants cultivated in India are …. . .
 a) Morus alba b) Morus indica
 c) Morus laevigata d) All of these

54. Which of the following is a type of shoot grafting?
 a) Whip grafting b) Wedge grafting
 c) Both a & b d) Flute budding

55. ….. mans the separation of bud along with bark from the stalk.
 a) Path budding b) T-budding
 c) Flute budding d) Whip grafting

56. Eggs of Bombyx mori are ….. Type.
 a) Centrolecithal b) Isolecithal
 c) Telolecithal d) Both a & c

57. In ….. yolk is present centrally.
 a) Isolecithal egg b) Telolecithal egg
 c) Centrolecithal egg d) Both a & b

58. The larva of Bombyx mori is …. . .
 a) Phytophagous b) Planktophagous
 c) Haematophagous d) Mycophagous

59. Eggs of silk moth are of ….. type.
 a) Megalecithal b) Centrolecithal
 c) Both a & b d) Isolecithal

60. ….. Is the silk covering of pupa.
 a) Endochorion b) Exochorion
 c) Cocoon d) Premantum

61. Pupa lives inside the …. . .
 a) Cuticle b) Cocoon
 c) Exochorion d) Reeling

62. ….. is the 3rd stage in the life cycle of silk moth.
    a) Pupa b) Cocoon
    c) Myx d) Chop
63. ….. is an appliance used to help the silkworms to spin the cocoon.
    a) Sechhi disc b) Centrifuge
    c) Mountage d) None of these
64. Mountage is also known as …… .
    a) Cocoonage b) Reeling
    c) Premantum d) Pupa
65. The mountages used in India are…… .
    a) Chandrika b) Netrica
    c) Paddy straw mountage d) All of these
66. Chandrika mountage is made up of …… .
    a) Bamboo mat b) Paddy straw
    c) Screens d) Polymer
67. ….. Is a type of mountage made of …… .
    a) Plastic b) Netlon polymer
    c) Paddy straw d) Bamboo mat
68. The silk reeling involves …… .
    a) Cocoon shfling b) Cocoon roddling
    c) Cocoon boiling d) All of these
69. For cocoon storage humidity should be maintained at …… .
    a) Above 70% b) Below 70%
    c) Between 80-120% d) 100%
70. ….. is a process of separation of cocoons according to their size.
    a) Cocoon riddling b) Cocoon deflossing
    c) Cocoon formation d) Both a & b
71. ….. means the unwinding of the silk filament from the cooked cocoons to get the silk.
    a) Canning b) Brushing
    c) Reeling d) Deflossing
72. Raw silk testing is done by …… .
    a) Visual tests b) Mechanical tests
    c) Both a & b d) Tavellate tests

73. The mechanical test for raw silk testing include ..... .
    a) Winding test b) Size test
    c) Seriplane test d) All of these
74. ..... is used to determine the evenness, cleanness and neatness defects of the silk.
    a) Seriplane test b) Winding test
    c) Tenacity and elongation test d) Cohesion test
75. The silkworm passes through ..... Instars.
    a) 2 b) 4
    c) 7 d) 5
76. The stages between the moults are known as ....
    a) Spinnert b) Instar
    c) Pinnaria d) Prolegs
77. ..... develops in to the pupa.
    a) 6th instar silkworm b) 5th instar silkworm
    c) 4th instar silkworm d) 7th instar silkworm
78. The cocoon is made up of a single long fibre in thread held together by ....
    a) Reeling b) X-mark
    c) Chiasma d) Seriein
79. What is the life span of an adult 'Bombyx mori'?
    a) 7 days b) 9 days
    c) 4 days d) 10 days
80. Silk contains a protein known as ....
    a) Fibroin b) Serioin
    c) Glyptin d) Both a & b
81. Silk gained entry to India through ....
    a) Russia b) Germany
    c) Tibet d) Australia
82. Biofertilizer that fixes nitrogen in Mulberry is...... .
    a) Azatobacter b) Rhizobium
    c) Catkin d) None of these

83. Suitable soil type for Mulberry is ..... .
    a) Rockey soil
    b) Limerich soil
    c) Porous soil
    d) Red loamy soil
84. Head quarter of National Silkworm Seed Organization is at ..... .
    a) Mysore
    b) Banglore
    c) Hyderabad
    d) Pune
85. Scientific name of Uzi fly is ..... .
    a) Exoristis bombycis
    b) Bombyx mori
    c) Bacillus proteus
    d) Spicaria prasina
86. Maximum infection of Uzi fly is recorded in ..... .
    a) Summer season
    b) Winter season
    c) Rainy season
    d) Both a & b
87. ..... is used for physical control of zi fly.
    a) CORTEA
    b) Uzicide tablet
    c) Simazine
    d) Endrin
88. Chemical control of Uzi fly is done by ..... .
    a) Spraying uzicide
    b) Spraying uzipowder
    c) Spraying 2% bleaching powder solution
    d) All of these
89. Symptoms of infections of Uzi fly includes ..... .
    a) Infected silkworms show black spots on their body surface
    b) Loss of weight due to continuous loss of fat bodies and their tissues
    c) Stop feeding
    d) All of these
90. Tasar silk is generated by the silkworm ..... .
    a) Antheraea mylitta
    b) Philosamia ricini
    c) Both a & b
    d) None of these
91. Eri silk is generated by of which silkworm?
    a) Antheraea mylitta
    b) Philosamia ricini
    c) Both a & b
    d) None of these

92. Types of silk based on species is …… .
    a) Mulberry
    b) Muga
    c) Eri
    d) All of these
93. …… . Is a disease of silkworms caused by Nosema bombycis.
    a) Chawkie rearing
    b) Pebrine
    c) Voltism
    d) Ecorace
94. Major constraints in sericulture are …… .
    a) Price fluctuation in cocoon
    b) Lack of education among the sericulture workers
    c) Absence of storage facilities
    d) All of these
95. Scientific name of muga silkworm is …… .
    a) Philosamia ricini
    b) Antheracea assama
    c) Bombyx mori
    d) None of these
96. Ericulture is mainly concentrated in …… .
    a) Assam
    b) Tripura
    c) West Bengal
    d) All of these
97. Main reasons for practicing sericulture are …… .
    a) High employment potential
    b) Women friendly occupation
    c) Important agro-based enterprise adding value in villages
    d) All of these
98. Mulberry accounted for …… . of the total raw silk production in the country.
    a) 40%
    b) 91%
    c) 28%
    d) 33%
99. Office of Indian silk Export promotion Council is at …… .
    a) New Delhi
    b) Bangalore
    c) Kolkata
    d) Chennai
100. Indian Silk Export Promotion Council was established in …… .
    a) 1980
    b) 1992
    c) 1983
    d) 1998

101. Maggot disease in silkworm is caused by ..... .

a) Nosema bombycis
b) Tricholyga pyocyones
c) Bacillus proteus
d) Bacillus sorbillans

102. The presence of milky white cylindrical eggs on the skin of silkworm larvae is a symptom of ....

a) Maggot disease
b) Pebrine disease
c) Green muscardine
d) Coila disease

103. Pebrine infection takes through ..... .

a) The mouth of larvae at the time of feeding
b) The mother's ovary
c) Both a & b
d) None of these

104. Types of polyhedrosis found in silkworms are ..... .

a) Nuclear polyhedrosis
b) Intestinal cytoplasmic polyhedrosis
c) Intestinal nuclear polyhedrosis
d) All of these

105. Counter measures for polyhedrosis in silkworms include ..... .

a) The diseased larvae should be removed
b) The high temperature, low temperature and high moisture should be avoided
c) The rearing rooms, tools and utensils should be cleaned and disinfected
d) All of these

106. Symptoms of intestinal cytoplasmic polyhedrosis are ..... .

a) Occurrence of translucent cephalothorax
b) Shrinkage of body size
c) Excretion of white coloured loose faeces by larvae
d) All of these

107. Symptoms of nuclear polyhedrosis are ..... .

a) Translucent cephalothorax
b) Body shrinkage
c) Diarrhoea
d) All of these

108. Symptoms of flacherie includes ....
   a) Loss of appetite
   b) Transversed cephalothorax
   c) Vomiting and diarrhea
   d) All of these
109. Counter measures for flacherie are ....
   a) Resistant silkworm races should be selected
   b) The rearing room, tools & utensils should be disinfected
   c) High quality mulberry leaves should be provided
   d) All of these
110. Bacteria attacking weak silkworms are ....
   a) Streptococci sp
   b) Coli aerogenous bacilli
   c) Proteus group bacilli
   d) All of these
111. Septi caemia is a ....
   a) Protozoan infection
   b) Viral infection
   c) Bacterial infection
   d) Worm infection
112. Depticaemia is caused by ....
   a) Bacillus megatherium
   b) Bacillus proteus
   c) Bacillus pyocyones
   d) All of these
113. .... Is a fungal disease of silkworms.
   a) Septicaemia
   b) Argulosis
   c) Green muscardine
   d) Flacherie
114. Scientific name of Green muscardine is ....
   a) Spicaria prasina
   b) Bacillus prodigious
   c) Costia necatrix
   d) Tricholyga sorbillans
115. National Sericulture Project was launched in ....
   a) 1979
   b) 1992
   c) 1989
   d) 1968
116. Oak Tasar silk Research Station is located at ....
   a) Imphal (Manipur)
   b) Batote (Jammu and Kashmir)
   c) Bhimtal (Uttar Pradesh
   d) All of these

117. Eri Silk Research Station is situated at ....
    a) Boko (Assam)
    b) Mendipathar (Meghalaya)
    c) Pampore (Kashmir)
    d) Ranchi (Bihar)
118. In Assam Mulberry silk Research Station is located at ....
    a) Majra
    b) Koraput
    c) Coonoor
    d) Jorhat
119. In West Bengal Mulberry silk Research is located at ....
    a) Mothabari
    b) Barrackpore
    c) Midhapore
    d) Koraput
120. Office of Central Silk Board is located at ....
    a) Mumbai
    b) Chennai
    c) Bangalore
    d) Pune

Answer keys

1	b)	Sericulture
2	d)	Government of India
3	b)	a=5, b=1, c=6, d=2, e=3, f=4
4	a)	2017-2020
5	a)	Vanya silks
6	c)	a=iii, b=v, c=vii, d=i, e=ii, f=iv, g=vii
7	d)	Both a & b
8	c)	Both a & b
9	a)	China
10	d)	Assam
11	d)	All of these
12	d)	Both a & b
13	b)	1989
14	c)	Five
15	d)	All of these
16	a)	Berhampore
17	c)	Bangalore
18	c)	1948
19	d)	All of these
20	a)	1972
21	c)	Mysore
22	a)	1983
23	d)	Mendipathor (Meghalaya)

24	d)	All of these
25	c)	Buko (Assam)
26	b)	China
27	a)	Its natural food is mulberry leaf
28	b)	Saturniidae
29	c)	Machilus plant
30	a)	Eri silkworm
31	b)	Uni-voltine
32	a)	Multi-voltine
33	b)	Internal
34	b)	Diapause
35	b)	Non-diapause eggs
36	c)	Caterpillar
37	c)	Both a & b
38	c)	4-6gm
39	a)	Cocoon
40	d)	All of these
41	a)	Moraceae
42	b)	Morus alba
43	c)	Morus indica
44	c)	Both a & b
45	d)	All of these
46	c)	Both a & b
47	b)	Stifling
48	d)	All of these
49	b)	Reeling
50	c)	Both a & b
51	a)	Silk
52	c)	Moriculture
53	d)	All of these
54	c)	Both a & b
55	a)	Path budding
56	a)	Centrolecithal
57	c)	Centrolecithal egg
58	a)	Phytophagous
59	c)	Both a & b
60	c)	Cocoon
61	b)	Cocoon
62	a)	Pupa

63	c)	Mountage
64	a)	Cocoonage
65	d)	All of these
66	a)	Bamboo mat
67	b)	Netlon polymer
68	d)	All of these
69	b)	Below 70%
70	a)	Cocoon riddling
71	c)	Reeling
72	c)	Both a & b
73	d)	All of these
74	a)	Seriplane test
75	d)	5
76	b)	Instar
77	b)	5th instar silkworm
78	d)	Seriein
79	c)	4 days
80	d)	Both a & b
81	c)	Tibet
82	a)	Azatobacter
83	d)	Red loamy soil
84	b)	Banglore
85	a)	Exoristis bombycis
86	c)	Rainy season
87	b)	Uzicide tablet
88	d)	All of these
89	d)	All of these
90	a)	Antheraea mylitta
91	b)	Philosamia ricini
92	d)	All of these
93	b)	Pebrine
94	d)	All of these
95	b)	Antheracea assama
96	d)	All of these
97	d)	All of these
98	b)	91%
99	a)	New Delhi
100	c)	1983
101	d)	Bacillus sorbillans

102	a)	Maggot disease
103	c)	Both a & b
104	d)	All of these
105	d)	All of these
106	d)	All of these
107	d)	All of these
108	d)	All of these
109	d)	All of these
110	d)	All of these
111	c)	Bacterial infection
112	d)	All of these
113	c)	Green muscardine
114	a)	Spicaria prasina
115	c)	1989
116	d)	All of these
117	b)	Mendipathar (Meghalaya)
118	d)	Jorhat
119	a)	Mothabari
120	c)	Bangalore

11

Vermitechnology

1. The rearing of earthworms for composting wastes into valuable nutrients is known as ....
   a) Organic farming b) Organic manuring
   c) Vermitechnology d) Sustainable technology
2. Vermitechnology provide ....
   a) Vermiwash b) Vermicompost
   c) Organic fertilizer d) All of these
3. Vermicompost improves ...... property of soil.
   a) Physical b) Chemical
   c) Biological d) All of these
4. In class oligochaeta ....
   a) Parapodia absent b) Clitellum present
   c) Pharynx without jaws d) All of these
5. In which order earthworm is placed?
   a) Oilgochaeta b) Frrantia
   c) Neooligochaeta d) Sedentaria
6. Earthworm is commonly known as ....
   a) Angler worm b) Dew worm
   c) Night crawler d) All of these
7. Earthworms are ....
   a) Nocturnal b) Detritivores
   c) Both a & b d) None of these
8. Earthworms are ....
   a) Bilaterally symmetrical b) Metamerically segmented
   c) Soft bodied d) All of these

9. Example of epigeic earthworm is ..... .
   a) Perionyx excavates b) Eisenia fetida
   c) Lumbricus rubellus d) All of these
10. Institute of National Organic Agriculture (INORa) islocated at ..... .
   a) Hyderabad b) Delhi
   c) Pune d) Patna
11. Example of endogeic earthworm is ..... .
   a) Megascolex konkanensis b) Polypheretima elongata
   c) Pontoscolex corethrurus d) All of these
12. Example of anecic earthworm is ..... .
   a) Lampito mauritii b) Lumbricus terrestris
   c) Polypheretima elongata d) All of these
13. Scientific name of Washington Giant earthworm is ..... .
   a) Drilolerius americanus b) Aporrectodea longa
   c) Lampito mauritii d) Lumbricus longa
14. Which of the following is known as Mega-earthworm diversity zone?
   a) West coast plain b) Western Ghats
   c) Both a & b d) Western dry region
15. ..... . is an example of chinese worm.
   a) Eisenia fetida b) Perionyx excavates
   c) Lumbricus terrtestris d) Both a & b
16. Eisenia fetida is commonly known as ....
   a) Brandling worm b) Tiger worm
   c) Red wiggler worm d) All of these
17. ..... . is an European night crawler.
   a) Eisenia hortensis b) Endrilus eugeniae
   c) Metaphire anomala d) Perionynx excavates
18. ..... . is commonly known as African night crawler.
   a) Eisenia hortensis b) Endrilus eugeniae
   c) Drawida willsi d) Metaphire posthuman
19. In India ..... . is suited for vermiculture.
   a) Eisenia fetida b) Perionyx pinarria
   c) Metaphire posthuman d) All of these

20. The mass of processed soil particles voided by earthworms is called ...
    a) Worm cast b) Vermiwash
    c) Breaker bar d) Vermibed
21. The Allyl isothiocynate (AITC) is used for ..... .
    a) Earthworm collection
    b) for treatment of earthworm diseases
    c) for purification of vermiwash
    d) For earthworm mating
22. Chemical used for earthworm collection is ..... .
    a) Formaldehyde b) Potassium permagnate
    c) Allyl-isothiocynate (AITC) d) All of these
23. Site for vermiculture should be ..... .
    a) Slope area b) Elevated place
    c) Both a & b d) None of these
24. Suitable species for vermiculture are ..... .
    a) Perionyx excavates b) Lampito mauritii
    c) Eisenia fetida d) All of these
25. ..... . is the substratum on which the earthworms live and multiply.
    a) Vermibed b) Vermicompost
    c) Arsel mat d) None of these
26. Material used for preparation of vermibed is ..... .
    a) Shredded paper waste b) Saw dust
    c) Weeds d) All of these
27. In which method epigeic and anecic species of earthworms are used jointly?
    a) Polyculture method b) Monoculture
    c) Integrated culture] d) Monosex culture
28. Earthworm feeds on ..... .
    a) Vegetable scraps b) Crop waste
    c) Grasses d) All of these
29. Suitable condition for earthworm is ..... .
    a) 75-90% moisture b) Adequate aeration
    c) 25-30% temperature d) All of these

30. Earthworms prefer ..... .
    a) Acidic pH b) Neutral pH
    c) Basic pH d) Both a & b
31. The optimum soil salinity for earthworm should be ..... .
    a) 0.5% b) 3.5%
    c) 8.2% d) 4.9%
32. The process of conversion of organic wastes into compost by earthworm is called ..... .
    a) Residue b) Guano
    c) Vermicomposting d) Vermiconversion
33. The conversion of organic wastes into soil additives by earthworms is known as ..... .
    a) Vermicomposting b) Vermiconversion
    c) Metaconversion d) Both a & b
34. .... is production of large quantities of desired worms for vermi-composting.
    a) Vermiculture b) Vermirocks
    c) Ver constriction d) Vermiwash
35. Method used for vermicomposting is ..... .
    a) Pit method b) Heap method
    c) Wedge system d) All of these
36. The vermicompost obtained from ..... . shows higher concentration of micronutrient and macronutrient.
    a) Monoculture method b) Polyculture method
    c) Large scale method d) Integrated method
37. Light retraction method is used for ..... .
    a) Earthworm harvesting b) Earthworm breeding
    c) Earthworm feeding d) Earthworm segregation
38. In Indoor vermicomposting the worms need..............
    a) Moisture b) Dark
    c) Food d) All of these
39. Vermicomposting in a limited volume of organic wastes is called ..... .
    a) Batch system b) Bar system
    c) Continuous system d) Parallel system

40. Batch system includes ....
    a) Windrow method b) Wedge method
    c) Stacked bins d) All of these

41. ...... is an automated system.
    a) Batch system b) Continuous flow system
    c) Both a & b d) None of these

42. Stacked bins method of vermicomposting was devised by ....
    a) George (2004) b) Goose (1875)
    c) Gillt (1998) d) Fritz (1990)

43. Vermicomposting in a walled enclosure constructed in a shed is called ....
    a) Heap method b) Wedge method
    c) Bed method d) Vermireactor method

44. Continuous flow system of vermicomposting was devised by ....
    a) Dr.Clive Edwards (1980) b) George (2004)
    c) Max Clients (1982) d) M.Bitch (2004)

45. ...... was modified by the Oregon soil corporation.
    a) Continuous flow system b) Beed method
    c) Stacked bins d) Windrow method

46. Vermitech-200 is a ....
    a) Continuous flow system b) Heap system
    c) Clinting system d) Both a & b

47. Vermitech-200 was designed by ....
    a) The Medical University of South Carolina
    b) Rothamstead Agricultural Research Station
    c) Indian Council of Agricultural Research
    d) Canvey International,China

48. Bhawalkar Ecological Research Institute is located in ....
    a) Bhopal b) Mumbai
    c) Pune d) Darjeeling

49. The mass of processed soil particles voided by earthworms is known as ....
    a) Worm cast b) Vermicasts
    c) Worm casting d) All of these

50. The worm casts are made in ....

a) Mouth of earthworm
b) Pharynx of earthworm
c) Anus of earthworm
d) Intestine of earthworm

51. Worm casts are ....

a) Good organic manure
b) Good vermicompost
c) Both a & b
d) None of these

52. The vermicompost contains ....

a) Organic carbon
b) Nitrogen
c) Zinc
d) All of these

53. ...... is a liquid extract collected after the passage of water through a mass of earthworms and cow dung.

a) Vermiwash
b) Vermicompost
c) Vermibed
d) Both a & b

54. ...... functions as a plant growth regulator.

a) Vermiwash
b) CIFAX
c) Vermix
d) Vermipromot

55. ...... contains enzymes, vitamins and hormones.

a) Vermibed
b) Vermix
c) Vermiwash
d) Both a & b

56. ...... is used as foliar spray.

a) Vermiwash
b) Vermifilter
c) Vermiguano
d) None of these

57. ...... is the natural vermiculture ecosystem used to treat effluent water.

a) Vermiwash
b) Vermix
c) Vermifilter
d) Vermixylo

58. Earthworms play a key role in soil biology by serving as ....

a) Bioreactors
b) Probiotics
c) Antibiotic
c) Protomix

59. ...... is glorified as the 'Father of Modern Organic Agriculture'.

a) Sir Albert Howard
b) Sir Albert Wutz
c) Francis gilto
d) P.Denmark

60. Vermicomposting produces ....

a) Earthworms b) Vermicast
c) Vermiwash d) All of these

61. ...... is the technology used to dispose chemical contaminants by earthworms.

a) Vermiremediation b) Vermidisposal technique
c) Vermiremedial separation d) Both a & b

62. ...... are the bio-indicators of heavy metal contamination in the soil.

a) Earthworms b) All worms
c) Tapeworms d) Hookworms

63. ...... are the earthworms living on the uppermost layers of the soil.

a) Endogeic b) Epigeic
c) Mesgeic d) Columngeic

64. A habitat earthworms burrow continuously to form a network of channels is known as ....

a) Epigeic b) Endogeic
c) Mesogeic d) Enzogeic

65. ...... is a litter dwelling earthworm.

a) Dendrobaena veneta b) Eisenia fetida
c) Eisenia Andrei d) All of these

66. ...... is an example of topsoil dwelling earthworm.

a) Allolobophora chlorotica b) Lumbricus terrestris
c) Eudrilus eugeniae d) Both a & b

67. ...... is a deep burrowing earthworm.

a) Lumbricus terrestris b) Allolobophora chlorotica
c) Both a & b d) None of these

68. Earthworms obtain their nutrition from ....

a) Fungi b) Nematodes
c) Both a & b d) Protozoans

69. Tunnel trap and electrical methods are used for ....

a) Earthworm collection b) Vermiwash preparation
c) Both a & b d) Earthworm identification

70. Vermimeal is used for ..... .

a) Birds b) Fish

c) Both a & b d) None of these

71. ..... . feeds deeper beneath the surface and ingest large quantities of organically rich soil.

a) Geophagous worms b) Phagous worms

c) Both a & b d) Sucker worms

72. Geophagous worms are known as ..... .

a) Humus feeders b) Liquid feeders

c) Soil feeders d) Solid feeders

73. Example of Geophagous worm is ..... .

a) Metaphire posthuman b) Octochetona thurstoni

c) Both a & b d) Eisenia fetida

74. Which of the following chemical is used for protecting vermi-bed from ants?

a) DDT b) Chloroform

c) Chlorpyriophosphate d) Acetic acid

75. Vermicompost is used as a biofertilizer because it is rich in ..... .

a) Nitrogen b) Calcium

c) Phosphorus d) All of these

76. Which of the following procedures are used by the farmers to multiply the earthworms?

a) By adding plant materials

b) By adding cow dung

c) By mixing more amount of biodegradable wastes

d) All of these

77. ..... . can not be used for vermicomposting.

a) Cow dung b) Plant materials

c) Animal wastes d) All of these

78. What type of bin is best for vermicomposting?

a) Plastic b) Wood

c) Metal d) All of these

79. …… . is known as black gold in vermicomposting.

a) Worm oil b) Pink worms

c) Worm casts d) Black oil

80. How long will it take for the worm population to double for vermiculture?

a) Three months b) Two months

c) Five months d) Six months

81. Which of the following are benefits of vermicomposting?

a) Improve soil texture b) Reduce erosion

c) Improve water holding capacity d) All of these

82. Vermiculture is also known as …… .

a) Earthworm farming b) Vermicomposting

c) Solid waste management d) None of these

83. Which is the best method to recycle agricultural and domestic waste.

a) Auqculture b) Sericulture

c) Apiculture d) Vermicomposting

84. Which state launched Godhan Nyay Yojana?

a) Maharashtra b) Punjab

c) Chhattisgarh d) Bihar

85. In India total agricultural land where vermicompost is practiced is …… .

a) 3.5 million hectares b) 7 million hectares

c) 2 million hectares d) 1.2 million hectares

86. Vermicompost is useful because …… .

a) It improves plant growth

b) It improves soil fertility

c) It protects plants from harsh climatic conditions

d) All of these

87. Vermicompost may be for …… .

a) Farming or agriculture applications

b) Landscaping applications

c) Compost tea

d) All of these

88. The recommended bedding ingredients for the vermicomposting bin are …… .
   a) Coconut coir b) Wood chips
   c) Aged compost d) All of these
89. Compost material created by worm is …… .
   a) Larger than 100 microns b) Smaller than 20 microns
   c) Smaller than 1 micron d) Larger than 500 microns
90. Harvesting time of compost range from …… .
   a) 10-12 months b) One year
   c) 3-6 months d) 10 days
91. Product cost of 1kg vermicompost is …… .
   a) 10-20 rupees b) 25-30 rupees
   c) 4-5 rupees d) More than 50 rupees
92. …… . enhances the process of decomposition of organic matter in soil.
   a) Coconut coir b) Vermicompost
   c) Woodchip d) All of these
93. …… . kills the earthworm.
   a) Water stresses b) pH at 7
   c) 75-90% moisture d) Adequate aeration
94. …… . act as biological pesticides and biological fertilizer.
   a) Vermiwash b) Vermicompost
   c) Both a & b d) Coconut coir
95. Which of the following is a detrivorous earthworm?
   a) Eisenia fetida b) Lampito mauritii
   c) Perionyx ecxavatus d) All of these
96. Octochaetona thurstoni is …… .
   a) Detritivorous b) Geophagous
   c) Planktivorous d) Microplanktivorous
97. Verimwash contains …… .
   a) Plant growth hormones b) Macronutrients
   c) Vitamins d) All of these
98. Vermicultutre technology is emerging as …… .
   a) Environmentally sustainable b) Economically viable
   c) Socially acceptable d) All of these

99. …… . is a humus like end product obtained by the action of earthworms on organic materials.

a) Mucus
b) Vermicompost
c) Both a & b
d) Spermatheca

100. Abiotic factors which affect vermicomposting are …… .

a) pH
b) Temperature
c) Moisture content
d) All of these

Answer Keys

1	c)	Vermitechnology
2	d)	All of these
3	d)	All of these
4	d)	All of these
5	c)	Neooligochaeta
6	d)	All of these
7	c)	Both a & b
8	d)	All of these
9	d)	All of these
10	c)	Pune
11	d)	All of these
12	d)	All of these
13	a)	*Drilolerius americanus*
14	c)	Both a & b
15	a)	Eisenia fetida
16	d)	All of these
17	a)	*Eisenia hortensis*
18	b)	*Endrilus eugeniae*
19	a)	Eisenia fetida
20	a)	Worm cast
21	a)	Earthworm collection
22	d)	All of these
23	c)	Both a & b
24	d)	All of these
25	a)	Vermibed
26	d)	All of these
27	a)	Polyculture method
28	d)	All of these
29	d)	All of these

10. ...... Identified as a good host for kusumi brood lac.
    a) Albizzia sp.
    b) Tachardia sp.
    c) Ribbania sp.
    d) Azardica sp.
11. The major constituent of lac is ....
    a) Wax
    b) Dye
    c) Resin
    d) Ash
12. The dusty lac eliminated by sieving is refuse lac known as ....
    a) Minamata
    b) Myleon
    c) Mytas
    d) Molamma
13. Available common hand-made seed lac grades in market are ....
    a) Fine by Sakhi
    b) Genuine by Sakhi
    c) Golden kusumi
    d) All of these
14. .... is the name of finished product and is commonly used across the world.
    a) Moltas
    b) Shellac
    c) Manbhum
    d) Both b & c
15. Common commercially available handmade grades of shellac are ......
    a) Lemon one shellac
    b) Lemon two shellac
    c) Superior kusumi lemon
    d) All of these
16. ...... is a mixture of anthro quinoid derivatives.
    a) Lemon shellac
    b) Waxy bleach
    c) Lac dye
    d) Molamma
17. ...... Is traditionally used to colour wool and silk.
    a) Lac dye
    b) Micromata
    c) Maytas
    d) Both b & c
18. Lac wax is a mixture of ....
    a) Higher alcohols
    b) Acids
    c) Easter
    d) All of these
19. Lac wax is used in ....
    a) Polishes applied on shoes and automobiles
    b) Food and drug tablet finishing
    c) Lipsticks and crayons
    d) All of these

20. …… . Is a natural gum resin.
   a) Mosoc b) Shellac
   c) Garnet d) Blonde
21. Shellac is …… . .
   a) Slightly heavier than water b) Softens at 65-70°C
   c) Insoluble in water d) All of these
22. Shellac is …… . .
   a) Thermoplastic b) UV-resistant
   c) Powerful bonding material d) All of these
23. Shellac is used in …… . .
   a) Fruit coatings
   b) Preparation of gramophone records
   c) Manufacturing of lithographic ink
   d) All of these
24. Bleached shellac is used in …… . .
   a) Binding b) Food packaging
   c) Fast drying d) All of these
25. Dewaxed bleached shellac is used in …… . .
   a) Coating of fruits
   b) Coating in cosmetics industry
   c) Coating in tablets and capsules
   d) All of these
26. …… . is obtained from shellac by saponification.
   a) Aleuritic acid b) Oxalic acid
   c) Picric acid d) Nitric acid
27. Aleuritic acid easters are used in preparation of …… . .
   a) Lacquers b) Fibres
   c) Plastics d) All of these
28. Lac insects are parasitized by small winged insect belonging to order …. . . .
   a) Chalcidae b) Hymenoptera
   c) Holoceraca d) Lepidoptera

29. …… is commonly known as 'Black lac moth'.
    a) Holocerca pulverea b) Chrysopa spp
    c) Eublemma amabilis d) Hymenoptera
30. …… known as 'Lac wing fly'.
    a) Eublemma spp b) Chrysopa spp
    c) Aleuritica spp d) Both a & b
31. Indian Lac Research Institute was established in ……
    a) 1984 b) 1963
    c) 1924 d) 1943
32. Indian Lac Research Institute is situated at ……
    a) Mumbai b) Kochi
    c) Ranchi d) Dehradun
33. India ranks …… In the production of lac.
    a) First b) Second
    c) Third d) Fourth
34. In India, most of the lac is produced in ……
    a) Chhatisgarh b) Jharkhand
    c) Both a & b d) Bihar
35. The first step in the lac cultivation is ……
    a) Swarming b) Ceupe culture
    c) Shell plantation d) Inoculation
36. …… is the process by which young ones get associated properly with the host plant.
    a) Inoculation b) Nutrition
    c) Swarming d) Clumping
37. Drawbacks of natural incubation are ……
    a) Incomplete nutrition
    b) Irregular inoculation
    c) Multiplication of parasites and predators
    d) All of these
38. Which are the types of Rangini crop ?
    a) Baisakhi b) Kartiki
    c) Both a & b d) Jethi

39. The types of kusumi crop is ....

a) Baisakhi b) Agahani
c) Jethi d) Both b & c

40. In ...... when one group of host plants is under the process of cultivation of lac, other groups of host plants would be under rest.

a) Alternation of plant b) Ceupe system
c) Both a & b d) Baisakhi

41. Common predators damaging lac cultivation are ....

a) Eublemma ambilis b) Holocera pulverea
c) Both a & b d) Erencyrtus dewitzii

42. The lac insects are parasitized by ....

a) Tetrastichus purpureus b) Erencyrtus dewitzii
c) Both a & b d) None of these

43. Abiotic enemies in lac culture are ....

a) High light intensity b) High temperature
c) High humidity d) All of these

44. ...... secretes lac throughout the life.

a) Male b) Female
c) Both a & b d) Youngones

45. ...... do not take major part in the secretion of lac.

a) Male b) Female
c) Both a & b d) Adult males

46. ...... has short life period.

a) Female b) Male
c) Both a & b d) Sterile individuals

47. In lac insect ...... Is larger and measures about 4 to 5mm in length.

a) Male b) Female
c) Nymph d) Both b & c

48. In lac insect mouth parts are of ...... type.

a) Piercing and sucking type b) Biting type
c) Chewing type d) Grinding type

49. How much eggs are produced by mature female lac insect?

a) 1000-1500 b) 1500-2000

c) 200-500 d) 50-100

50. The oviposition takes place in ....

a) Incubating chamber b) Ovary

c) Testis d) Resinous cell

Answer Keys

1	a)	J. Kerr (1782)
2	a)	Tachardia lacca
3	c)	Protection from predators
4	b)	Six months
5	d)	All of these
6	c)	Ovoviviparous
7	a)	Crawlers
8	d)	All of these
9	a)	Kusumi lac
10	a)	Albizzia sp.
11	c)	Resin
12	d)	Molamma
13	d)	All of these
14	c)	Manbhum
15	d)	All of these
16	c)	Lac dye
17	a)	Lac dye
18	d)	All of these
19	d)	All of these
20	b)	Shellac
21	d)	All of these
22	d)	All of these
23	d)	All of these
24	d)	All of these
25	d)	All of these
26	a)	Aleuritic acid
27	d)	All of these
28	b)	Hymenoptera
29	a)	Holocerca pulverea
30	b)	Chrysopa spp

31	c)	1924
32	c)	Ranchi
33	a)	First
34	c)	Both a & b
35	d)	Inoculation
36	a)	Inoculation
37	d)	All of these
38	c)	Both a & b
39	d)	Both b & c
40	b)	Ceupe system
41	c)	Both a & b
42	c)	Both a & b
43	d)	All of these
44	b)	Female
45	a)	Male
46	b)	Male
47	b)	Female
48	a)	Piercing and sucking type
49	c)	200-500
50	a)	Incubating chamber